Daniel Dejica, Gyde Hansen, Peter Sandrini, Iulia Para (eds.)

Language in the Digital Era. Challenges and Perspectives

Daniel Dejica, Gyde Hansen,
Peter Sandrini, Iulia Para (eds.)

Language in the Digital Era

Challenges and Perspectives

DE GRUYTER
OPEN

ISBN: 978-3-11-047204-2
e-ISBN: 978-3-11-047205-9

Published by De Gruyter Open Ltd, Warsaw/Berlin
Part of Walter de Gruyter GmbH, Berlin/Boston
The book is published with open access at www.degruyter.com.

Library of Congress Cataloging-in-Publication Data
A CIP catalog record for this book has been applied for at the Library of Congress.

www.degruyteropen.com

Cover illustration: © Hilch

Contents

Part III: **Language Teaching and Learning in the Age of Technology**

Daniel Dejica, Gyde Hansen, Peter Sandrini, Iulia Para (eds.)
Introduction

This collected volume brings together the contributions of several humanities scholars who focus on the evolution of language in the digital era. The eighteen contributions are divided into three thematic parts, which explore general aspects of humanities and linguistics in the digital environment, the evolution of language and translation in today's digitized society, and the changes, challenges and perspectives of language teaching and learning in the age of technology.

Part I, *Humanities Gone Digital*, explores general aspects of humanities and linguistics in the digital environment.

In the opening chapter, *Recent Trends in Digital Humanities Scholarship*, Mary P. Sheridan highlights the increasing role of digital media within higher education. The author claims that digital technologies are changing the ways we learn and teach, as well as the ways we compose and research. According to her, these changes are occurring throughout the academy, including the humanities—a set of disciplines less associated with technology. Mary P. Sheridan describes the rise of the Digital Humanities (DH) in the United States, defines and illustrates DH projects from many countries, and offers suggestions for incorporating DH projects in our work.

In the second chapter, *Theme-Rheme Analysis of English and Romanian Tourism Websites*, Claudia Elena Stoian and Daniel Dejica present the results of a contrastive Theme-Rheme analysis performed on a corpus of commercial websites from Great Britain and Romania, meant to promote these countries and some of their heritage sites internationally via the Internet. Using a framework provided by Systemic Functional Linguistics, the authors identify, analyze and compare the Themes and the Thematic structures prevalent in these websites.

In the third chapter, *Necessary and Luxury English Loanwords in Some Romanian Online Newspapers and Magazines*, Simona Șimon claims that the socio-economic and political context of contemporary Romania favours the private and professional communication between the local people and other nationalities. Since English is the most frequently taught language in the Romanian public schools, it is no wonder that it is often used in private and professional exchanges. The author claims that an expected consequence of this situation is the borrowing of English words into the Romanian language. In her study, she identifies the necessary and luxury English loanwords used in some Romanian online newspapers and magazines, and presents her own conclusion in this respect.

Corpus Linguistics Outcomes and Applications in the Digital Era by Diana Oțăt is the last chapter of the first part of this book. The author addresses specific aspects with regard to current trends and applications in corpora design and corpora compiling in the digital era. She focuses on novel interdisciplinary approaches to language study via corpora investigation and applied computer-assisted analysis tools. Corpus linguistics is approached in her study in relation to real-life communicative contexts, i.e. authentic language inputs that facilitate descriptive and functional language research, and, more particularly specialised languages in an attempt to provide reliable solutions in compliance with the dynamics of contemporary linguistics. The author pays special attention to quantitative corpora investigation methods and modern information-extracting technologies as applied to different dimensions of language study. The state-of-the-art outline of corpus linguistics that she provides aims at highlighting further research directions in interconnected fields such as contrastive linguistics, sociolinguistics, lexicography, translation studies and foreign language teaching.

Part II focuses on language and translation and includes topics that discuss the digital translation policy, new technologies and specialised translation, online resources for terminology management, translation of online advertising, or subtitling.

In *Towards a Digital Translation Policy*, Peter Sandrini, purports that translation policy represents a core component of an efficient language policy, and that intrinsically, it guarantees that multilingual communication works as intended within a company, organization or institution. The author looks at the contents of such a translation policy and outlines the effects of a general digitalization and globalization of the translation industry. By describing these changes, he stresses the necessity of a sensible translation technology policy – in the sense of what kind of translation technology should be deployed, used by whom and when, on what kind of texts, etc. Finally yet importantly, Peter Sandrini addresses the need for an overall strategy with regard to translation data and language resources in general, and integrates it into the concept of an overall translation technology policy.

In the second chapter of this part, *The Impact of New Technologies on Specialised Translation*, Mariana Pitar claims that new technologies have become a useful tool for all science fields, whether they are exact sciences or humanistic sciences. The author highlights the contribution of technology on specialized translation and presents an overview of the translation tools used during various stages of the translation process, which include text analysers, online specialised dictionaries, specialised terminological resources, translation memories, or translation verifying tools.

The Transfer of Signs between Heterogeneous Systems: Incongruent Equivalences is the title of the next challenging chapter, in which Felix Nicolau tries to provide answers to some daring questions. The author discusses about the stance of Translation Studies today, wonders whether it has passed the linguistic limits still proclaimed by a semiotician like Umberto Eco, and asks himself whether it would not be the time to access more courageously the intersemiotic interregnum with its heterogeneous transfer of signs, as it happens in advertising, concrete poetry and stage or filmed version of famous texts. Other provoking insights focus on the role of translation and translators in a post-industrial society, which blends globalized edutainment and corporatist efficiency, prejudice, reverse colonialism and anti-establishment movements, and on the types of equivalence are we supposed to choose in order to persuade today.

In chapter four, *Evaluating Online Resources for Terminology Management in Legal Translation*, Titela Vîlceanu focuses on the evaluation, both quantitative and qualitative, of relevant and reliable online resources – bilingual and multilingual glossaries, multilingual databases, monolingual and bilingual dictionaries – from a twofold perspective: to enhance internal and external coherence with respect to terminology management, and to secure error-free legal translation. The language pair envisaged by the author is English-Romanian since the Romanian legal translation market has developed considerably after 2007. The author claims that this market is characterised by variability with respect to the use of legal terminology, even within the same text type, she aims to identify the causes of this variability and she suggests ways of improving/securing the quality of legal translation via a criterion-referenced evaluation of available online resources.

In *To Delete or to Add? Omissions and Additions in Two Romanian Translations of Jack and the Beanstalk*, Loredana Pungă investigates the effect that omissions and additions have on the propositional, expressive and evoked meaning in two Romanian translations of the classical English fairy tale *Jack and the beanstalk* (a printed and an electronic text) and highlights the consequences of these translation/adaptation techniques on how the Romanian texts are received by their readers.

In chapter six, *A Standards-Based Contrastive Analysis of Online and Printed Technical Translations in Romanian*, Daniel Dejica creates a standards-based grid, which he uses to analyse the formal and content-specific aspects of 45 interdependent translations in printed and electronic formats. His research offered the opportunity to reach several conclusions useful to translation professionals and relevant for the bodies responsible with the maintenance of translation standards, and to raise a series of challenging questions for translation researchers or scholars.

The last two chapters of Part 2 focus on subtitling. In *Extratextual Elements in Subtitling – The Battle of Linguistic and Cultural Codes*, Violeta Tănase points out the constraints and challenges brought up by extratextual elements, especially when it comes to the translation of culture-bound terms. Such elements which add up to the 'foreignness' of an audiovisual product and have an impact upon the subtitler's choices include objects, settings, physiognomies, gestures, costumes, music and noise tracks, background conversation, radio announcements and television commercials, intertitles, canned laughter and written materials such as posters, billboards or newspapers.

Elena Laura Vulpoiu claims there are few monographs and studies on the realization of the subtitling process, and she considers necessary to bring to light some rules and guidelines used in this type of audiovisual translation. In her contribution, *Subtitling in Romania and Spain: A Contrastive Analysis*, the author investigates and describes the practice and characteristics of subtitling in Romania and compares them with the standards used in Spain. Her study is based on empirical data provided by a translator of the Romanian national television (TVR 1).

The six chapters in **Part III**, *Language Teaching and Learning in the Age of Technology*, focus on language teaching and learning and address the changes, challenges and perspectives of didactics in the age of technology.

László Komlósi's research findings, presented in *Digital Literacy and the Challenges of Digital Technologies for Learning*, indicate that unprecedented development and innovation in information and communication technologies exert unforeseen impact on social cognition, information processing and human learning. The author acknowledges a paradigm change from linear information processing based on narrative mental structures and cultural conceptualizations to parallel and connected network-based information processing making use of fragmented, encapsulated information chunks provided by a plethora of information sources. In opposition to traditional learning conditions which have been constituted by a hierarchically-determined accumulation and distribution of knowledge and information with norm-based behavior patterns, new digital cognition and information management involve greater complexities and fragmented narratives connected in a non-linear, non-deterministic distribution design in the environment of augmented realities. László Komlósi points out the innovative drive which resides in interactive informational frameworks that consist of a multitude of connections of the Connected Cognitive Entities (CCEs) to other cognitive entities and create novel patterns of learning styles and learning habits.

In the second chapter of this part, *On the Use of Hypermediality in Teaching Culture in a German as a Foreign Language Context*, Karla Lupșan shows the importance of the use of hypermedia (i.e. the combination of text elements, graphics, video and audio) in teaching culture in a German as a foreign language context and presents a didactical concept of an innovative use of parallel texts as tools for self-directed knowledge acquisition.

In her contribution, *Online Communication – Netspeak. The Internet as a Facilitator for New ways of Communication and the Impact on Our Language*, Iulia Para analyses online communication and the language of the Internet. She identifies the main features of *netspeak* and describes what makes it different from standard language. She also presents some of the most commonly used abbreviations, especially in texting, a very popular means of online communication nowadays.

The aim of Alexandra Jic's chapter, *Young English Learners in the Digital Age*, is to discuss the importance of employing modern technologies to support English language learning of young learners. The author starts from outlining the need to reform teaching styles in order to meet the needs of the 21st century young learners who have grown up in a digital world. Then, she moves on to examine the need for a reformed educational system in Romania within the context of the newly introduced digital books for the 1st and 2nd grades. The author also presents some of the problems that might occur when using technologies and makes some suggestions that may enable teachers to overcome these possible issues and be able to use technology to the benefit of the learners. Alexandra Jic bases her claims and suggestions on the findings of a survey carried out in several Romanian schools at the end of January 2015 among teachers and young learners regarding their opinion on the use of technology in classroom and the biggest challenges to use digital books in school.

In the next chapter, *Training and Development in the Digital Era*, Simona Olaru-Poşiar looks at the steps of the training cycle and examines the management of training and development in the digital era. The author claims that in today's digital era, we develop better with the help of smart devices, meant to ease our path towards knowledge and that our continuous education or continuous training is the purpose for our development as human beings. We develop with the help of smart devices and are influenced by them whether we like it or not, because our environment is today conditioned by the World Wide Web. The author concludes that this "web of information" marks our development and the way we communicate.

In the last chapter of the book, *Developing Communication Skills in Romania in the Digital Era*, Valentina Mureşan explores different problematic issues of postmodern teaching in Romania, starting from a necessary change of the local teachers' mindset in order to meet the expectations of the new generation of digital learners in order to "engage" them and help them develop as good communicators. In this context she discusses the faulty understanding of the concept of communicative competence as a unique skill, rather than a multifaceted one, a fact that can be explained by looking at co-existing patterns of old and new methodologies, approaches and techniques. The author claims that although the demands of post-modern language teaching in Romania seem to have changed, there is still a certain resistance to it, linguistic proficiency still being the focus of many EFL teachers.

The book is recommended to scholars, professionals, students and anyone interested in the changes within the humanities in conjunction with technological innovation or in the ways language is adapting to the challenges of today's digitized world.

Notes on the editorial board and contributors

Daniel Dejica, PhD, is an associate professor in translation studies at *Politehnica* University of Timisoara, Romania. His research interests include translation theory and methodology, LSP translation, and discourse analysis for translation purposes. Daniel Dejica is a member of the Advanced Translation Research Center (ATRC) team at the University of Saarbrücken, Germany and a member of the Doctoral Studies Committee of the European Society for Translation Studies (EST). He has been co-editing the *Proceedings of the Professional Communication and Translation Studies* conference, organized at Politehnica University of Timişoara since 2001; he is also a member in the editorial boards of other international peer-reviewed journals including *conneXions: international professional communication journal* (New Mexico Tech), *MuTra Journal* (University of Saarbrücken), or *The European English Messenger* (ESSE - European Society for the Study of English). Between 2012 and 2015 Daniel Dejica had been the Head of the Department of Communication and Foreign Languages at Politehnica University of Timisoara; he is currently the Director of the Interlingua Language Center at the same university. E-mail: daniel.dejica@upt.ro

Gyde Hansen, Prof. dr. habil. and **PhD** teaches at the Copenhagen Business School (CBS) in the disciplines: linguistics, comparative grammar, intercultural communication and negotiation, textual analysis, text revision, semiotics and marketing, translation processes and translation theory, and philosophy of science. Her empirical research projects are the *TRAP-project* (Translation processes), the *Copenhagen Retrospection Project*, a longitudinal study, *From Student to Expert*, and a project on translation revision processes, *TraREdit*. E-mail: gh.ibc@cbs.dk and the website: www.gydehansen.dk

Alexandra Jic, PhD, has taught English at the West University of Timisoara. Her main research interests include literature, lexis and teacher training. She has published several articles in these fields and she is a member of The European Society for the Study of English and of the Romanian Society for English and American Studies. E-mail: alexandra_jic@ yahoo.com

László I. Komlósi is a professor of organizational communication and argumentation studies with a research background in cognitive pragmatics, cultural linguistics, formal philosophy and argumentation theory at the Széchenyi István University in Győr, Hungary. In his professional career, he has taught at several universities in Hungary and abroad, has been an active participant in and frequent keynote speaker on international conferences. He has conducted research with the help of Hungarian and international grants and has published extensively over the years. He has supervised a dozen of PhD dissertations. Professor Komlósi is a member of several international academic societies and is editorial board member of a number of international journals. E-mail: komlosi.laszlo@sze.hu

Karla Lupșan, PhD, is currently an assistant professor at the Faculty of Letters, History and Theology, West University of Timișoara. Her main areas of research are German linguistics, contrastive grammar (German-Romanian), translation theory and practice, and translation methodology. She is the author of several books and scientific articles in these fields and member of the editorial board of international peer-reviewed journals, such as the journal for German studies *Temeswarer Beiträge zur Germansitik* or the journal for translation and traductology *Translationes*. E-mail: lupsan_karla@yahoo.com

Valentina Mureșan has a PhD from the West University of Timisoara in English Applied Linguistics. She is currently working as an assistant at the Department of English, of the West University of Timisoara, Romania. Her research interests and published papers have focused on various aspects of applied linguistics, language teaching methodology, discourse analysis and translation studies. E-mail: muresancarina1@ gmail.com

Felix Nicolau is Professor in the Department of Foreign Languages and Communication, The Technical University of Civil Engineering, Bucharest, Romania. He defended his PhD in Comparative Literature in 2003 and is the author of five volumes of poetry (*Kamceatka – time IS honey*, 2014), two novels and eight books of literary and communication theory: *Take the Floor. Professional Communication Theoretically Contextualized* (2014), *Cultural Communication: Approaches to Modernity and Postmodernity* (2014), *Comunicare şi creativitate. Interpretarea textului contemporan* (*Communication and Creativity. The Interpretation of Contemporary Text*, 2014), *Homo Imprudens* (2006), *Anticanonice* (*Anticanonicals*, 2009), *Codul lui Eminescu* (*Eminescu's Code*, 2010), and *Estetica inumană: de la Postmodernism la Facebook* (*The Inhuman Aesthetics: from Postmodernism to Facebook*, 2013). He is a member in the editorial boards of "Poesis International", "The Muse – an International Journal of Poetry" and "Metaliteratura" magazines. His areas of interest are translation studies, the theory of communication, comparative literature, cultural studies, translation studies, and British and American studies. E-mail: hamsun10@yahoo.com

Simona Olaru-Poşiar, PhD, is an assistant at Victor Babes University of Medicine and Pharmacy in Timisoara, where she teaches English and German. She authored several books, including *Motivul nebuniei în literatura germane – de la Georg Büchner la Patrick Süskind* [*The motif of madness in German literature – from Georg Büchner to Patrick Süskind*] and *Literatur und Medizin. Wahn und Wahnsinn in der Medizinwissenschaft und in der Literatur* [*Literature and Medicine. Illusion and madness in medical science and literature*]. Simona Olaru-Poşiar is a member of several national associations including the Gesellschaft der Germanisten Rumäniens and CPR- Romanian College of Psychology. Email address: simona.posiar@yahoo.com

Diana Oţăt, PhD, is an assistant in the Department of British-American and German Studies, Faculty of Letters, University of Craiova. Her teaching and research activity focuses on Translation Studies, Legal English, Pragmatics and Intercultural communication. She is a certified International ECL Examiner (Romanian, English and German) in accordance with the Common European Framework of Reference for Languages, and a Foreign Language Trainer, certified by the Ministry of Labour, Family and Social Protection. She has also attended national and international conferences and workshops. She has authored a course book (*Insights and functional models in translation theory and practice*) and about 10 scientific articles. E-mail: otatdiana@gmail.com

Iulia Para, PhD, is an assistant professor in English at the Faculty of Economics and Business Administration, West University of Timisoara, Romania, where she teaches English as the Language of Communication in Business. She is the author and co-author of several books, including *Anglicisms in the Romanian Business Vocabulary: On Analysis in the Finance and Banking Texts in the Ziarul Financiar* (published at LAP LAMBERT Academic Publishing, Germany) or *An Introduction to Communication in English* co-authored with Judith Moise and Daniela Istodor, and of 28 articles published in Romanian and international journals. Iulia Para is a member of the PGV scientific international network; reviewer for WSEAS International Conferences and TJEB journal; coordinator of the International Relations Office of the Faculty of Economics and Business Administration and member of the CSR team of the same faculty. E-mail: iulia.para@e-uvt.ro

Mariana Pitar, PhD, is an associate professor in translation studies at the Romance Languages Department, Faculty of Letters, History and Theology, West University of Timişoara, Romania. She teaches terminology, translation of audio-visual documents, computer-assisted translation (CAT), discourse analysis. She published several articles on linguistics, specialized translation or terminology, and the books: *Textul injonctiv. Repere teoretice* (2014) [*Injunctive Text. Theoretical Landmarks*], *Genurile textului injonctiv* (2007) [*The Genres of Injunctive Text*], and *Manual de terminologie şi terminografie* (2013) (*A Book of Terminology and Terminography*). E-mail: pitarmariana@yahoo.fr

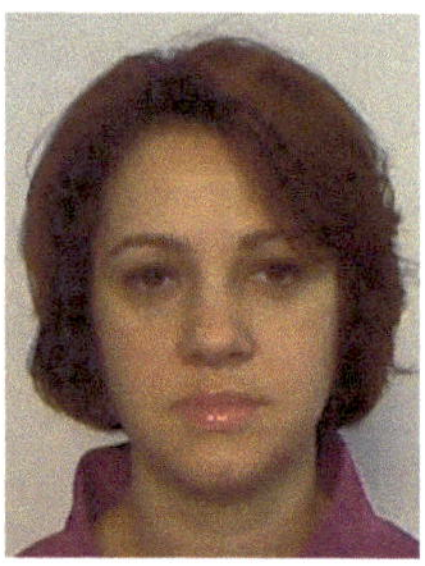

Loredana Pungă is an associate professor at the West University pf Timișoara, Romania. Her domains of expertise are English lexicology, applied and cognitive linguistics and translation studies. Her publications include *On Language and Ecology* (Timișoara, Editura Universității de Vest, 2006); *English for Students of Kinetotherapy* (Timișoara, Editura ArtPress, 2007, co-author Carmen Nedelea); *English for Sports and Games* (Timișoara, Editura Universității de Vest, 2010, co-author Carmen Nedelea); *Words about Words. An Introduction to English Lexicology* (Timișoara, Editura Universității de Vest, 2011). Loredana Pungă is co-editor of *Language in Use. The Case of Youth Entertainment Magazines* (Newcastle, Cambridge Scholars Publishing, 2010) and *A Journey through Knowledge. Festschrift in Honour of Hortensia Pârlog* (Newcastle, Cambridge Scholars Publishimg, 2012). She is also co-editor of *BAS. British and American Studies and Translationes,* both yearly journals governed by the Department of Foreign Languages at the university where she teaches. She has published articles in her areas of research both in Romania and abroad. E-mail: loredana.punga@ yahoo.ro

Peter Sandrini, PhD, has earned a degree in translation studies; his doctoral dissertation focused on legal terminology. He has published extensively on legal terminology, translation, website localization and translation technology. He is currently attached to the Department of Translation Studies at the University of Innsbruck as an Assistant Professor where he lectures on Translation Technology, Terminology as well as LSP Translation. He is also the initiator of USBTrans and tuxtrans, a project which aims at bringing Open Source Software to translators training and the translator. E-mail: peter.sandrini@uibk.ac.at

Mary P. Sheridan, PhD, is Professor of English at the University of Louisville, in the United States. Sheridan writes and teaches on questions relating to digital composing, community engagement, and feminist methodologies. She has written *Girls, Feminism, and Grassroots Literacies: Activism in the* GirlZone, co-authored *Design Literacies: Learning and Innovation in the Digital Age*, and co-edited *Writing Studies Research in Practice: Methods and Methodologies* as well as *Feminism and Composition: A Critical Sourcebook.* Her articles have appeared in *CCC, Computers & Composition, Kairos, JAC, Written Communication, Feminist Teacher, Composition Studies,* and *Journal of Basic Writing.* Sheridan won the 2010 Winifred Bryan Horner Outstanding Book Award from Coalition of Women Scholars in the History of Rhetoric and Composition; the 2009 Civic Scholarship/ Book of the Year Award from *Reflections: A Journal of Writing, Service-Learning, and Community Literacy*; and, as part of a collaborative group, the *Computer and Composition*'s Michelle Kendrick Outstanding Digital Production/Scholarship Award for 2008. E-mail: maryp.sheridan@louisville.edu

Claudia Elena Stoian, PhD, is a graduate in English Philology from the Autonomous University of Madrid, Spain. She holds a PhD in Applied Linguistics from the same university. She has published several papers on discourse analysis and cultural differences. At present, she is teaching Translation, Discourse Analysis and English as a Foreign Language at the *Politehnica* University of Timisoara, Romania, and working as a translator of English, Spanish and Romanian. E-mail: claudia.stoian@upt.ro

Simona Şimon, PhD, is an assistant professor in the Department of Communication and Foreign Languages of the *Politehnica* University of Timişoara, where she teaches Interpreting, Pragmatics, Oral Communication Skills in English, Notions of Publicity. She holds a doctoral degree in Philology from the University of the West, Timişoara. She is the author of the book *The Persuasive Function of Written Advertisements* and of several scientific articles. Her research interests are in the field of applied linguistics, genre studies, interpreting, advertising and teaching. E-mail: simona.simon@upt.ro

Violeta Tănase is a PhD student at *Alexandru Ioan Cuza* University of Iasi, Romania. She holds a Master's degree in Translation studies from the same university and is currently working for her PhD thesis on the topic of "Subtitling humor as a form of intercultural communication". Research interests: audiovisual translation, the translation of humor as a form of intercultural communication, humor production and reception mechanisms in stand-up comedy, the subtitling of extralinguistic and intralinguistic culture-bound references in sitcoms and teen soaps, translation strategies for rendering taboo language in audiovisual translation. E-mail: elena.tanase@uaic.ro

Titela Vîlceanu, PhD, is an associate professor in the Department of British-American and German Studies, Faculty of Letters; Director of The Department of Publications and Media, University of Craiova. Her research focus is on Translation Studies, Pragmatics Intercultural communication and Legal English. She is also a methodologist accredited by the British Council Romania and by the University of Edinburgh, Institute for Applied Language Studies, an ARACIS (Romanian Agency for Quality Assurance in Higher Education) evaluator in the field of Applied Modern Languages, a Romanian language linguistic administrator (AD5) in the field of translation, certified by the European Personnel Selection Office (EPSO)-European Commission, and Bologna expert certified by EACEA. She has attended numerous international conferences in Romania and abroad and authored about 50 articles and 5 books. She has participated in 12 international interdisciplinary research projects. E-mail: elavilceanu@yahoo.com.

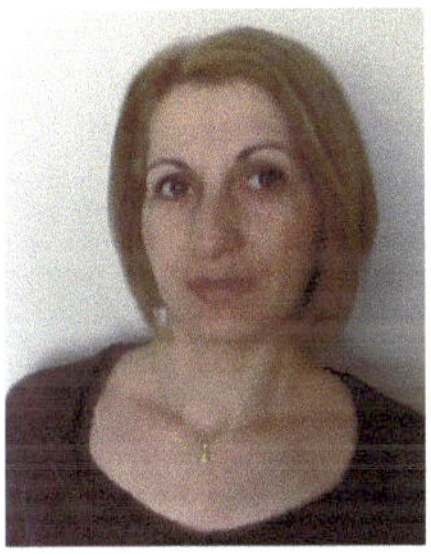

Elena Laura Vulpoiu is doing her PhD studies in Applied Languages, Literature and Translation (Descriptive, empirical and experimental studies of audiovisual translations) at the University Jaume I, Spain. Her thesis is specifically aimed at the theoretical and applied analysis in the process of subtitling and audiovisual translation in English and Romanian language. Her first contributions in this area have been published in several scientific journals and she has participated in international and national conferences and seminars. E-mail: lauravulpoiu@gmail.com

Part I: **Humanities Gone Digital**

Mary P. Sheridan

1 Recent Trends in Digital Humanities Scholarship

1.1 Introduction: Concerns and Possibilities That Give Rise to the Digital Humanities

Education is changing, and digital work is both reflective of and instrumental for these changes. Yet, whereas digital technology has long been central in many academic arenas, it has received mixed reviews by those in the humanities. The past decade, however, has introduced several factors that have been giving technology a more prominent role in these ostensibly less-technical forums.

One factor has been an anxiety that US higher education is not preparing students for today's changing workforce. As US jobs continue to be off-shored due to global market conditions or automated due to technological advances, many have asked if higher education is appropriately preparing US students to be future workers and leaders in an economy that demands innovation and entrepreneurialism for quick adaptations to global systems in flux. This anxiety about student preparation, coupled with rising student debt, has led citizens, businesses, government officials and policy makers[1] to question what students are receiving for their increasingly expensive higher education. More, such questions have nudged academic institutions, programs, and departments to explore how globally networked digital media can keep universities relevant in today's changing contexts.

A second, related concern that has helped fuel increased interest in technology relates to funding for higher education. For years there has been an ongoing withdrawal of US governmental funding, with 48 of the 50 states continuing to cut financial support (Center for Budget and Policy Priorities, 2014) to the point where some argue that the government will all but de-fund these universities within the coming decades. As universities scramble to fill that budgetary gap, they rely on a series of measures, including the controversial courting of business sponsorships.

1 An earlier high-profile example is when, in 2006, Margaret Spellings, the then chair of the US Department of Education under George W. Bush, convened a Commission on the Future of Higher Education working group to recommend future educational trajectories. The resulting report, "A Test of Leadership, Charting the Future of US Higher Education," started by asking what businesses and citizens were getting for the high price tag of education. Illustrating the wide-spread anxiety about whether US higher education is preparing students, and therefore the US, to be leaders in today's globalized world, this report asked educators to justify how it was preparing today's workers, a justification those in higher education continue to be asked to make today (Commission Appointed by Secretary of Education Margaret Spellings, 2006).

University athletic divisions have solicited and maintained corporate alliances for years, but expanding that reliance, critics fear, makes universities beholden to these companies beyond their ubiquitous advertising at university sporting events or their names on campus buildings, facilities, or academic positions. Donations that have explicit strings attached (e.g., about research agendas, curricular choices, hiring practices) can compromise the intellectual freedom that universities champion,[2] yet, as long as these economic conditions remain, cash-strapped universities have few options for funding and will, therefore, need to negotiate new sponsorships and the incumbent responsibilities that go along with such relationships.

Among the many concerns facing higher education, these two—anxiety about student preparation and concerns about sponsors who fund and shape that preparation—show higher education's twinned problems of needing to do more with less. In that difficult spot, universities are learning to do differently, and technology is increasingly touted as a means both to prepare students for the digitally saturated employment world of today and to develop partnerships beyond the traditional ones. Not surprisingly, many in higher education have pinned their hopes for emerging educational practices on the use of digital media throughout higher education, including the humanities. Fortunately, this shift is becoming easier in the classroom, in part because of the world in which we live today.

According to the Pew Charitable Trust, digital media is ubiquitous in many students' lives (Lenhart, *et al.*, 2008), and teachers across the university are tapping this interest. For example, the annual *Educause* list of the top 100 technologies teachers report using in their classrooms indicates that teachers are bringing high-end and everyday technology into their classrooms (Dahlstrom and Brooks, 2014; Purcell, *et al.*, 2013), and they are using that technology in various ways, whether to teach rhetorical strategies with social media, efficient data management for research projects, or effective collaboration in digitally mediated forums. This interest in using digital communication in the classroom is shared by university presidents as well; when asked what higher education will look like in the year 2020, the majority of university presidents responded that higher education will be quite different, with more digitally mediated teaching and learning than exists today (Anderson, et al., 2012). In fact, most university presidents anticipate a significant rise in online teaching

2 Consider, for instance, the controversy in the United States surrounding the Koch brothers donating large sums of money to Schools of Business, something that has occurred at my university as well. These two conservative donors often attach strings to their donations, such as a required reading for students or a review of academic candidates by the donors. While these Koch brother restrictions have been softened, in some instances this overall practice worries academics due to the perceived and/or actual effects that businesses—often with strong political agendas—may be buying higher education and its influence.

and learning, with half of those responding that digitally mediated teaching will be important to how a majority of students receive their education (Taylor, *et al.*, 2011).[3]

This general interest in and acceptance of digital technology in academia today is also evident in parts of the humanities. As technology finds a stronger foothold in the humanities, a conventionally less-technical arena than other parts of universities, more resources are becoming available to illustrate and explore the possibilities of the Digital Humanities, including: the US government sponsored National Endowment for the Humanities has a special section called the Office of the Digital Humanities; established book series have Digital Humanities collections, such as Blackwell's *A Companion to Digital Humanities* (Schreibman, Siemens, and Unsworth, 2004); and, academic conferences, disciplinary specializations, and academic jobs are devoted solely to the Digital Humanities.

Such markers highlight that the digital and indeed the Digital Humanities have arrived, yet the actual definition of the Digital Humanities is less stabilized. Many see that the term can push scholars to explore possibilities beyond what has been traditional inquiry. Others contest the term altogether; noting that the digital is infused in the sciences, engineering, and medicine (areas that do not use the "digital" modifier), this group pushes scholars to focus on defining the humanities in contemporary contexts. Still others use the term Digital Humanities to mean just about anything that uses digital media. Despite the diversity of views, the Digital Humanities generally refers to a set of methods and projects that investigate how the pairing of the terms "digital" and "humanities" extends one another. In other words, definitions coalesce around investigations into what it means to be human in the digitally networked information age. As such, these definitions tap traditional humanities strengths—such as synthesis, analysis, creation, and curation of artifacts with social/cultural significance—within digital networks to respond to cultural, economic, and global changes (Burdick, *et al.*, 2003, p. 82).

In sharing their work, Digital Humanities scholars not only build upon but also extend humanities knowledge, making their arguments in both traditional, longer academic forms (e.g., academic papers, research reports) and in innovative, shorter forms (e.g., data visualizations). These scholars also pursue novel methods that, among other things, open new forms of scholarly inquiry, assess and organize knowledge in different ways, and facilitate collaboration across both traditional disciplinary silos as well as academic and non-academic participants. While scholars have been asking new questions, providing inventive assessments and collaborating

3 Oft-cited examples include ways to provide less expensive and more inclusive educational options. For example, supporters of digitally mediated learning argue that technology can provide education more cheaply (e.g., online classes do not require expensive buildings or maintenance of physical classrooms) and entice more students (e.g., online classes allow working or rural students to take classes that may be impossible if classes require face-to-face class meeting).

for years, the degree to which these practices are the norm in Digital Humanities scholarship is distinctive. It is still too early to definitively state what exactly can come in this context, but already Digital Humanities scholars are charting several key paths, as the following examples illustrate.

1.2 Examples of Digital Humanities Projects

There are many ways to describe the diverse Digital Humanities projects that infuse long-standing traditions of humanistic inquiry into rapidly changing technological possibilities of today. Below are examples of three types: the *digitization of information*, which makes information widely and often freely available to a far greater range of stakeholders; the *born digital* projects, which call for multimodal, often interactive and immersive, engagements afforded by digitally mediated technologies; and, *the creation of tools* that allow for new ways to research and represent that research. Fostering collaboration and opening the possibility of innovative methods and questions, each of these projects extends humanities research by encouraging new ways of creating and representing knowledge.

1.2.1 Digitization of Existing Materials

One type of Digital Humanities project is the *digitization of existing materials*, which makes historically static material easily available and accessible to anyone with an Internet connection. While not an entirely new practice—scholars have had access to archives in the past—increasing access to this information radically changes the scope and possibilities of such work. Consider, for instance, Devon Elliott and William J. Turkel's use of early 20th-Century periodicals for stage magicians as way to develop image processing techniques that extract, classify, and visualize imagery. They did so both to better understand practices and changes in the field of magic, and to suggest ways that similar image mining processes might help scholars efficiently identify and analyze themes across digitized visual collections. Just as individual scholars may find the digitization of existing materials of value, nations and global groups can, too, as evident in the Woodrow Wilson Center Archive making available recently declassified documents on various themes, such as The Cold War, The Korean Conflict, and Nuclear History. Providing de-classified global communications, this free, digitally accessible archive changes who can easily do research, from scholars and public intellectuals to curious global citizens. The European Commission's *Digital Agenda For Europe: A Europe 2020 Initiative* also seeks to expand who can participate in important conversations, in this case by "improv[ing] the framework conditions for digitization and digital preservation" of key cultural material (n.p.). In short, the mass availability of shared documents allows for new degrees of openness and of collaboration among interested groups, regardless of their participants' status.

Digitizing data clearly has its rewards, and as more people, organizations and nations become interested in the possibilities, we will see more information and a greater range of people engaging with this information. Still, there are challenges that go beyond simply converting materials into .pdfs and .jpegs and posting them online. Researchers continue to struggle with finding and accessing texts that may exist in various formats (articles, reports, diaries, newspaper clippings, photos, recordings), securing or clearing copyrights that may be held by groups with competing agendas, digitizing materials for archival purposes, coding files for organization and retrieval, managing and maintaining archives in ways that others can readily access, and providing the means to expand these systems as more information becomes available. Such efforts require expertise, time, and money, and the demands for these resources will only grow as the frameworks and infrastructures supporting the resulting datasets become more complex. But the work is already being done, and the results at individual and global levels are promising. In this way, what may have started as an important but incremental change in who has access to data sets—a difference in *degree* of access—is developing into the coordination and mining of large scale, cross-disciplinary data sets that can change the types of research questions able to be explored—a difference in the *kind* of research easily accomplished in the Digital Humanities.

1.2.2 Born Digital Scholarship

A second type of Digital Humanities project is *born digital scholarship.* Rather than making traditional data available in a static format, born digital scholarship attempts to foster projects conceived entirely in digitally-mediated spaces.

A relatively early example is *Pox and the City: Edinburgh: A Digital Role Playing Game for the History of Medicine*, a virtual reality game developed by scholars in multiple countries (Canada, United States, United Kingdom) in multiple disciplines (epidemiology, computer sciences, education). Seeking to teach the complex, systems thinking needed for learning medicine, this immersive role-playing game asks students to become different characters, each of which attempts to address the spread of Small Pox in the city of Edinburgh around 1800. In their adopted roles, students use clues and primary historical documents provided for them in the game to learn how the advancement of science is influenced by factors as diverse as architecture, cultural and religious norms, and propaganda from rival medical professionals (Kean, 2013). While, in theory, students could imagine such a world and could themselves gather primary and secondary paper sources, *Pox and the City* was imagined and created for digital contexts. Specifically, by tapping the affordances of digital technologies, this project was conceived as a series of simulations that help students engage in the complex deductive and inductive reasoning needed to learn epidemiology (see Pox and the City).

A second, more recent example of born digital scholarship is The Virtual Reconstruction of an Afghan Refugee Camp as a Site of Cultural Memory, another project that seeks to recreate lived experience, this time an Afghan refugee center that arose in 1979, after the Soviet invasion of Afghanistan. Led by scholars from politics, architectural history, and performance studies, and working with local community organizations that have ties to the United Nations High Commissioner for Refugees, The Virtual Reconstruction project uses actual materials from the camp (e.g., letters, journals, diaries, photos, audio tapes, videos) in order to develop "online multi-player games to create a virtual community of testimony, witness, recovery and social bonding" (Burdick, *et al.*, 2013, p. 68-69). The project uses these primary sources to create an immersive environment for all, but perhaps especially for those displaced by the Afghan conflict. Examining "patterns of diaspora, assimilation, and cultural memory" (Burdick, *et al.*, 2003, p. 68), this project uses the textual and embodied testimony and performances from the camp inhabitants as a way to see the camp as both a "historical site" and as a "living memory" where history, diaspora, and trauma can be addressed.

In addition to exploring possibilities the digital offers, these and other examples of born digital scholarship reveal methodological dilemmas researchers face, perhaps especially when they need to negotiate potentially competing cultural values and ethical considerations. As these Digital Humanities scholars attempt to understand and represent how large-scale historical forces shape groups of people, they also must account for their own research choices by probing the consequences individual people may face if their personal stories are exposed. Such projects ask traditional humanities questions, such as: How do we determine which people and objects should be included and how should they be contextualized? Who benefits when historical materials are made available? And, what are the risks in such a project? These Digital Humanities projects also intensify more traditional humanities questions, such as: What happens when we change the scale of access to someone's story? How do we keep from being distracted by the medium (i.e., how do digital tools alter the way we convey a story)? How do we decide what work should be digitized? These decidedly humanistic inquiries are placed within contemporary digital frameworks that both extend and rework humanistic traditions, thus highlighting what Digital Humanities scholarship can contribute to such traditions.

1.2.3 Creation of Digital Tools

A third type of Digital Humanities scholarship, and one I find particularly important at the moment, is the *creation of digital tools* that gather, analyze, synthesize, and present the findings of scholarship in ways not readily done, or even possible, in more traditional scholarship.

One example from translation studies examines how core religious, literary, and cultural texts "translate" into other languages. For example, the Version Variation Visualization (VVV) project sponsored by the United Kingdom's Arts and Humanities Research Council is developing digital tools for creating, curating, exploring, and analyzing corpora of differing versions of texts. Led by an interdisciplinary team that includes researchers in the Departments of Computer Science, English, History, Classics, as well as Languages, Translation and Communication, the VVV looks at historical and contemporary retranslations of cross-culturally significant works of literature, scripture, and philosophy so as to develop two things: 1) new ways of understanding cultural change and diversity in and through such text corpora, and 2) new ways of communicating these understandings to both academic and lay communities, using data visualizations (see Delighted Beauty). An illustration of this work is the Multilingual crowd-sourcing of Shakespeare's *Othello* project, which seeks "to make digital tools to help us explore world culture, by comparing how the same work is translated differently, over time and space, in the same and different languages" (Delighted Beauty, n.p.). In doing so, the VVV aims to "develop text analysis and data visualization tools which will contribute to cross-cultural understandings and enable new research, new learning and teaching" (Delighted Beauty, n.p.). By mapping the shared and divergent understandings of certain translations, the resulting visualization (Figure 1.1) can show how shared texts do and do not translate in different cultures, indicating potential cultural misalignments.

A second example, "Networks in History: Data-Driven Tools for Analyzing Relationships Across Time," similarly seeks to create visualization tools, in this case a suite of tools that can help researchers interpret, analyze, and represent given *and missing* data from the massive influx of large, online data sets that are becoming increasingly available. Tapping the collaborative potential of international (Italian and US) and interdisciplinary scholars (from disciplines including History, French, Italian, English, Classics, and Media Technology, as well as other collaborative areas such as technology support, libraries, and research labs), "Networks in History" emerged out of ongoing projects that sought to examine the "impact that social networks have on our own professional lives [and research projects], as well as the power that digital technologies possess for revealing the organization of these networks" (NEH Office of Digital Humanities Implementation Grant, 2013, p. 3). Considering that these data sets are often historically incomplete, humanistic interpretive frameworks (a qualitative project) can help researchers make sense of what mistakenly seems a question of running-the-numbers on these data sets (a quantitative project) (NEH Office of Digital Humanities Implementation Grant, 2013, p. 4). As these examples illustrate, international, interdisciplinary collaboration can create new possibilities, ones that extend traditional humanities scholarship by altering the methods, teams, and circulation/representation of conventional humanities research.

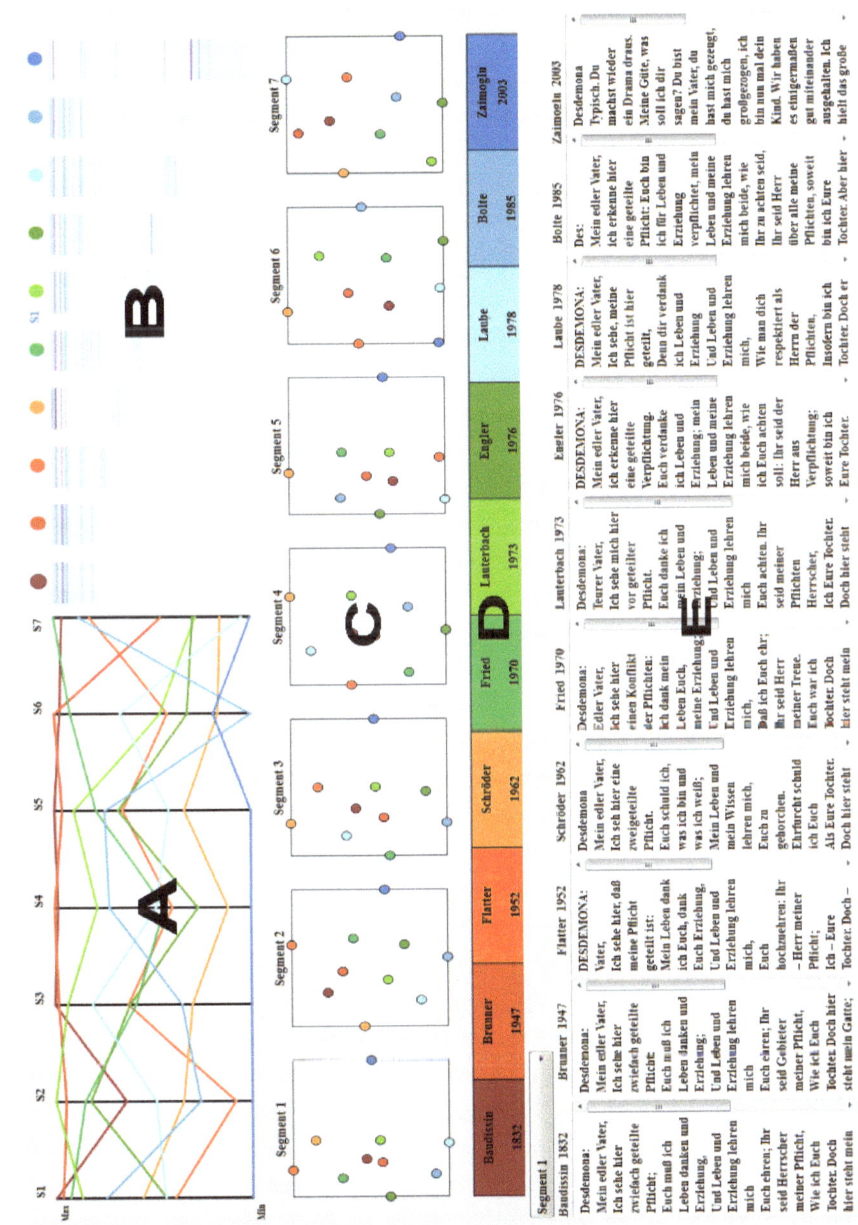

Figure 1.1: Visual Analysis of Segment Variation of German Translations of *Othello*. Visualization developed by Geng and Laramee (see Cheesman, Flanagan, & Thiel, 2013).

The above examples also illustrate that the distinction between types of Digital Humanities scholarship is often fuzzy; digitizing existing materials encourages researchers to ask new questions that require the development of new tools, which can encourage new types of digital scholarship. For example, exploring historical approaches to making sense of the millions of web-based images generated each day, Ian Milligan has used Web "scrapes" to generate data sets within millions of images to explore methods of "distant reading," what he calls "snapshots" of archived web content, to examine what large-scale archives as systems can help us see about the past. Similarly, Kate Bagnall and Tim Sherratt's *Invisible Australians: Living Under the White Australia Policy* project uses a facial detection script to extract images from exemption certificates, part of the archives borne out of Australia's Immigration Restriction Act 1901; compiling the resulting images into an infinite scroll, the authors allow viewers to see the *real* faces of so-called "White Australia." Regardless of the type of Digital Humanities scholarship, these examples illustrate how scholars in a variety of contexts and through a variety of methods are re-imagining how to conduct and represent their research.

1.3 Initial Steps for Incorporating DH Scholarship

Each of the above examples illustrates the interdisciplinary and collaborative potential of Digital Humanities projects. And yet, few of us have such resources readily available to us. In fact, the projects above are so large it may be hard to imagine how to start such work. Consequently, below is one final example—a substantially scaled down example—to demonstrate how even the most reticent can begin to incorporate digital technology in their work.

At the 2014 Thomas R. Watson Conference at the University of Louisville, we created a digital archive that invited conference participants to submit a 1 minute video of what it means to them to be a responsive teacher, researcher or community-engaged partner in our field today (Watson Archive). The various responses demonstrate a range of Digital Humanities scholarship. For example, most participants offered short video clips of them talking into a camera. Such responses are akin to the digitization of existing materials, in which the participant made widely available their response to the theme of the conference, something that could have been written out but something that can be far more accessible as a digital project. Other participants (e.g., DeVoss, Hesford, Sheridan) submitted videos echoing born-digital scholarship. Incorporating digital affordances such as images, sound, animation and embedded clips, these videos forwarded arguments in ways that exceed print-only possibilities. We hope this widely accessible digital archive can be a sandbox for those in our field to mine. Like those working with the VVV *Othello* project, people working with the Watson Digital Archive can create tools and visualizations that analyze and represent the data in ways that show, among other things, current narrative tropes about disciplinary self-representations. Collectively, these Watson Conference videos—now archived as a digital resource—can provide a low barrier for many to enter Digital Humanities scholarship, whether creating their own videos or analyzing others' videos. In this way, the Watson Digital archive functions as an invitation for people to explore and participate in Digital Humanities scholarship based on their own varying degrees of familiarity with the Digital Humanities itself.

1.4 Conclusions: Thoughts on Beginning Digital Humanities Projects

The Digital Humanities are emerging within contexts where concerns about higher education are being met by people's aspirations for technological possibilities. Among the many concerns, perhaps most notably concerns about who will fund today's educational projects, two questions are central: 1. what is higher education supposed to be doing? and, 2. who gets to decide that? The pervasiveness of digital media offers hope for both of these concerns, and the Digital Humanities in particular can help us pursue creative, critical, ethical, and analytical cultural projects within contemporary contexts.

Even as the above examples highlight the promise of Digital Humanities work, resources are needed to realize that promise. Take just one example: although funders (e.g., the National Endowment for the Humanities and the National Institute of Health in the US) encourage projects that draw upon interdisciplinary teams and that share their work with wider audiences, these goals are more likely achieved within a linked infrastructure that provides ways to share data and to collaborate across diverse networks. Such an infrastructure, however, raises important questions: Will universities support (with time and technological infrastructure) teachers and students to innovate and collaborate meaningfully, especially in hiring and promotion decisions, even if such work takes time and resources away from more traditional scholarship? Who will have access to these data/what types of surveillance and privacy will be in place? How will we avoid being overwhelmed by large amounts of data we can now easily collect?

Although these questions do not yet have satisfactory answers, diving into the Digital Humanities still seems well advised: it is not only our future but, as the research illustrates, it is very much part of our present. Consequently, I end this essay with three suggestions for how to embark on Digital Humanities projects. First, start small: Explore a visualization tool in your research, incorporate a digital assignment in your teaching, participate in a digitally mediated community of scholars exploring issues you care about. Then assess, revise, and try it again. Once you have accomplished this project, try something new.

Second, find one or more partners. Initially, a partner can be anyone interested in exploring similar questions, whether those questions concern a topic, a method, a pedagogy, or a combination thereof. Eventually, develop a team of people (from your home institution or from across the globe), likely humanists, designers, and technologists, who may develop larger projects with each contributing their own expertise to the endeavor.

Finally, know that success never happens the first time around. Indeed, failure is part of the process, but the next iteration is usually better. With continued practice—and on-going tweaking or redesigning—questions, projects, and contributions provide lenses for imagining and investigating questions, often in innovative ways. When we can accept this process-oriented approach toward research, teaching and learning, and when we can be open to unexpected possibilities along the way, we can see the potential the Digital Humanities have to offer our own work.

References

Anderson, J., Boyles, J. L., & Raine, L. (2012, July 27). Main findings: higher education's destination by 2020. Pew Research Center. Retrieved from http://www.pewinternet.org/2012/07/27/main-findings-higher-educations-destination-by-2020/ (accessed February 2015)

Bagnall, K. & Sherratt, T. Invisible Australians: living under the white Australia policy. Retrieved from http://invisibleaustralians.org (accessed August 2015)

Burdick, A., Drucker, J., Lunenfield, P., Presner, T., & Schnapp, J. (2013). Digital_Humanities. Cambridge, MA: MIT Press.

Center for Budget and Policy Priorities. (2014, May 1). State funding for higher education remains far below pre-recession levels in most states: changes in state spending per student, inflation adjusted, FY08-FY14. Center for Budget and Policy Priorities. Retrieved from http://www.cbpp.org/research/states-are-still-funding-higher-education-below-pre-recession-levels (accessed February 2015)

Cheesman, T., Flanagan, K., & Thiel, S. (2013, January). Translation array prototype 1: Project overview. Delighted Beauty. Retrieved from http://www.delightedbeauty.org/vvv/ (accessed 30 September, 2015)

Commission Appointed by Secretary of Education Margaret Spellings. (2006). A test of leadership: charting the future of US education. Washington, DC: Education Publications of the US Government.

Dahlstrom, E., & Brooks, D. C., with a foreword by D. Oblinger. (2014, July). ECAR study of faculty and information technology, 2014. Louisville, CO: ECAR. Retrieved from http://www.educause.edu/ecar (accessed February 2015)

Delighted Beauty. Retrieved from https://sites.google.com/site/delightedbeautyws/ (accessed January 2015)

Elliot, D., & Turkel, W. J. Faster than the eye: using computer vision to see magic. In K. Kee (Ed.), Seeing the past: augmented reality and computer vision to explore sources in the history of stage magic (unpublished manuscript).

European Commission. Digital agenda for Europe: A Europe 2020 initiative. Retrieved from http://ec.europa.eu/digital-agenda/en/digitisation-digital-preservation (accessed July 2015)

Kean, S. (2013, January/February). Pox in the city: from cows to controversy, the smallpox vaccine triumphs. Humanities 34(1). Retrieved from http://www.neh.gov/humanities/2013/january-february/feature/pox-in-the-city (accessed March 2013)

Lenhart, A., Arafeh, S., Smith, A., & MacGill, A. (2008, April 24). The lives of teens and their technology. Pew Research Center. Retrieved from http://www.pewinternet.org/2008/04/24/the-lives-of-teens-and-their-technology/ (accessed March 2009)

Milligan, I. Exploring web archives through tens of thousands of images." In K. Kee (Ed.), Seeing the past: augmented reality and computer vision to explore Sources in the history of stage magic (unpublished manuscript).

NEH Office of Digital Humanities Implementation Grant. (2013). Networks in history: data-driven tools for analyzing relationships across time. Retrieved from http://www.neh.gov/files/grants/stanford_university_data-driven_tools_for_analyzing_relationships_across_time.pdf (accessed July 2015)

Pox and the City: Edinburgh: A Digital Role-Playing Game for the History of Medicine. Retrieved from http://poxandthecity.blogspot.com (accessed January 2015)

Purcell, K., Heaps, A., Buchanan, J. & Friedrich, L. (2013, February 28). How teachers are using technology at home and in their classrooms. Pew Research Center. Retrieved from http://www.pewinternet.org/2013/02/28/how-teachers-are-using-technology-at-home-and-in-their-classrooms/ (accessed November 2014)

Schreibman, S., Siemens, R. G., & Unsworth, J. (2004). A companion to digital humanities. Malden, Mass.: Blackwell Publishing.

Taylor, P., Parker, K., Lenhart, A., & Patten, E. (2011, August 28). The digital revolution and higher education: college presidents, public differ on value of online learning. Pew Internet & American Life Project. Retrieved from http://www.pewinternet.org/files/old-media//Files/Reports/2011/PIP-Online-Learning.pdf (accessed March 2013)

Watson, Thomas R., Digital Archive. Retrieved from https://vimeo.com/watsonconference (accessed October 2014)

Woodrow Wilson Center Digital Archive: International History Declassified. Retrieved from http://digitalarchive.wilsoncenter.org (accessed February 2015)

Claudia Elena Stoian, Daniel Dejica

2 Theme–Rheme Analysis of English and Romanian Tourism Websites

2.1 Introduction

Promotion is essential for the tourism sector, as the product is not a regular one. In this case, the product is a service and has the following features: "intangible, inseparable, variable and perishable" (Morgan and Pritchard, 2000: 10). Tourism promotion has the purpose of configuring the tourist image of a destination while it points out its attractiveness (Calvi, 2006). A potential tourist decides to go on a holiday and becomes a real tourist "purely on the basis of symbolic expectations established promotionally through words, pictures, sounds and so forth" (Morgan and Pritchard, 2000: 10).

The present study looks at tourist promotional messages from a textual perspective. It presents a small-scale comparison of commercial tourist websites belonging to two different countries: Great Britain and Romania. These websites promote numerous landmarks belonging to their country and at the same time, offer users the possibility to buy a tour to/ticket for the landmark(s) promoted. The focus of the study is on the webpages used to promote these countries and their heritage sites internationally via the Internet, in particular, on their texts. The following World Heritage Sites have been chosen: Canterbury Cathedral, the Tower of London and Edinburgh, for Great Britain; and the Monastery of Horezu, the Dacian Fortresses of the Orastie Mountains and Sighisoara, for Romania.

The aim of the study is twofold: 1) to analyse the texts by identifying their Themes and Thematic structure and 2) to compare the results by looking at the similarities and differences that appear between the ways in which the commercial tourist texts are organized in light of their Themes and Thematic structure. The frameworks considered for the analyses are Systemic Functional Linguistics (Halliday, 1985, 1994; Halliday and Matthiessen, 2004) and Thematic patterns (Daneš, 1974).

2.1.1 Theoretical framework: The system of Theme

The creation of meaning in linguistic discourse can be explained and analysed by means of Systemic Functional Linguistics, which proposed the theory of a three-layered purpose of language. It is claimed that, like any semiotic system, language fulfils three communicative functions at the same time. These are ideational, i.e., the way reality is constructed in a discourse; interpersonal, i.e., the grammatical choices that enable people to enact their complex and diverse interpersonal relations; and textual, i.e., the internal organization and communicative nature of a text (Halliday and Matthiessen, 2004: 5). The last function is concerned with the creation of text into a meaningful whole and "enables the clause to be packed in ways which make it effective given its purpose and its context" (Eggins, 1994: 273). The system of Theme represents one of these ways.

Following Dejica (2005, 2009), Theme accounts for the coherence and internal organization of discourse by organizing the initiation of the clause and/or by directing the attention of the receiver of the message to the parts the sender wishes to emphasize. As such, the Thematic organisation of a clause is the most important element in the development of a text. The Theme choices guide the recipient throughout the unfolding message.

The system of Theme presents the clause as falling into just two main constituents: Theme, i.e., the first experiential constituent, which serves as a starting point for the message and tells what the clause is going to be about, and Rheme, i.e., the part of the clause in which the Theme is developed (Halliday and Matthiessen, 2004: 64-7; Eggins, 2004: 273-5). Theme and Rheme are frequently associated with given and new information, considering that "we typically depart from the familiar to head towards the unfamiliar" (Eggins, 2004: 275). As such, Theme is considered to express familiar or given information, in other words, information which has already been mentioned somewhere in the text or is familiar from the context. As opposed, Rheme presents unfamiliar or new information (Halliday, 1994: 59). Nevertheless, this may not always be the case because speakers or writers can interchange these associations for particular reasons (Lombardi Vallauri, 1995: 359).

The Theme element can be further classified according to its composition and status. Theme can be simple, when expressed by one element – the experiential one; and multiple, when expressed by more than just the experiential element. The other element(s) forming the Theme can be textual – show how clauses are linked together and give cohesion to the text, and/or interpersonal – signal the writer's point of view on the message (Thompson, 1996: 156-160; Halliday and Matthiessen, 2004: 68, 79-87). As far as status is concerned, its choice reflects the writer's purposes and depends on whether the clause is declarative or not. Theme can be unmarked, i.e., the typical or usual choice, for example, Subjects in declarative clauses, and marked, i.e., the atypical or unusual choice, like adjuncts in declarative clauses (Eggins, 1994: 296-7; Halliday and Matthiessen, 2004: 73-9). Unmarked Themes seem to carry the lowest degree of new information, while marked Themes tend to bring in new information, being both topically and informationally salient (Westergaard, 1986 cited in Manoliu-Manea, 1994: 230). However, according to McCabe (1999: 85), markedness can only be determined in context; in other words, the degree of markedness can vary (Thompson, 1996: 144-6).

2.1.1.1 Thematic structure

According to Dejica (2010: 22), individual Themes per se or their selection and identification in clauses are "not particularly significant", but their "overall choices and ordering" (Dejica 2004: 49) are. The selection of Themes is important as it organises the text and provides a point of orientation, revealing its underlying coherence and its method of development. Themes can develop following different Thematic patterns (Daneš, 1974), such as:

a) Theme re-interation or continuous pattern – the same Theme is kept over several clauses. In this case, the message shows a clear focus.
b) Thematic shifting – the Theme of one clause is taken from the Rheme of a previous clause. Thematic progression can be achieved by two different patterns. In the first one, zig-zag or linear pattern, part of the Rheme of each clause becomes the Theme of the following clause. Considering that it builds on newly introduced information, this development becomes cumulative. As for the second pattern, i.e., multiple-Theme or split, the Rheme of one clause introduces a number of different pieces of information, each of which is then picked up and made Theme in subsequent clauses. This last pattern provides frequently the underlying organising principle for a text (Eggins, 1994: 302-5).

Thematic development (Fries, 1983), thus, has several functions. According to Thompson (1996: 165-73), it points out the maintenance or progression of what the text is about at a certain point, specifies or changes the framework for the interpretation of the following clause(s), signals the boundaries of sections in the text, and shows what the speaker considers to be the useful or important starting point for her/his message.

The present study aims to analyse the Themes and Thematic structure of the chosen texts. The analysis can show their texture and organisation. It can also reveal the way the copywriter made clear her/his underlying concerns and priorities (Whittaker, 1995a, 1995b; Halliday and Matthiessen, 2004: 105).

2.2 Data selection and methodology

The study focuses on two commercial tourist websites: one British and the other Romanian. A number of search criteria have been followed in the selection of the corpus. English has been set as the language of promotion, since the study focuses on international promotion. Different types of well-known World Heritage Sites, i.e., religious, historical and urban landmarks from different regions, have been selected in order to cover possible diversity in the landmarks' features and their promotion. In addition, the analysis of more than one item per website aims to obtain a broader view and observe whether there is any distinctive promotional feature related to the site. The selection process has used the Google search engine and has taken the results generated by the word 'visit', followed by the name of the religious landmark. The first tour operator to appear has been checked also for the historical and urban landmarks. The tour operator has been chosen only if it promoted all the three landmarks.

The two commercial websites are Evan Evans Tours (www.evanevanstours.co.uk) and Ciao Romania (www.ciaoromania.co.uk). They are owned by tour operators that work in the countries mentioned and offer tours to visitors. They both promote landmarks and offer the possibility to buy tickets and tours online. In the case of every website, three webpages have been selected, each introducing a different national World Heritage Site, as mentioned above. The corpus of the study may be characterised briefly as follows: the field is tourism, particularly World Heritage Sites (religious, historic and urban landmarks); the medium is the Internet; the mode is the text with a length of approximately 95 words – British set and 55 words – Romanian set (the texts presenting urban landmarks are slightly longer); the language is English; the communicative purpose is to inform and persuade; the sender of the message is a tour operator; and the addressee is the general international public, who may be interested in gathering information, planning a trip and/or buying tours or travel packages.

The webpages advertise tours visiting different landmarks, which are not the object of discussion in this paper. As such, only the texts referring to the particular World Heritage Sites chosen have been analysed. We used Halliday's model (1985, 1994; Halliday and Matthiessen, 2004) for the analysis of Themes and Rhemes, and we also considered Daneš' classification of Thematic patterns (1974). The textual analysis has looked only at the independent conjoinable clause complexes, i.e., complexes that "consist of an independent clause together with all hypotactically related clauses and words that are dependent on that independent clause" (Fries 1995), because they seem to be the optimal method for showing Thematic progression (Thompson, 1996: 166; McCabe, 1999: 73). For each unit of analysis, the initial part of the clause up to the first experiential constituent was considered Theme. In the case of fronted dependent clauses, these were taken as Theme, as the point of departure for the whole complex. The analysis includes the types of Themes, their status, composition and their Thematic progression. For clarity reasons, a table presenting these aspects is included for each text.

2.3 Results of the Theme-Rheme analysis

The results of the analyses are presented in relation to the representation of the textual function for each text. The results are first presented and then compared.

2.3.1 British set of texts

As already mentioned, the texts considered for the British set are those presenting Canterbury Cathedral, the Tower of London and Edinburgh.

2.3.1.1 Canterbury Cathedral

There is no webpage dedicated entirely to Canterbury Cathedral; the landmark is part of a tour, together with other landmarks in the region of Kent. The Cathedral is presented mainly in relation to its importance for the Anglican faith.

The textual analysis of the text is presented in Table 2.1. (Please note that *T* in italics indicates a new Theme; pointed dashes are used to signal an unclear, but possible Thematic development; underlined words indicate textual Theme and double square brackets stand for embedded clauses).

Table 2.1: Canterbury Cathedral: Theme and Thematic structure

Theme	Type of Theme	Rheme	Thematic pattern
T1: Travelling through Kent, the county known as the 'Garden of England'	Dependent clause - Marked	R1: we make our way to Canterbury to visit *the Cathedral* (THEME FROM DEP. CL.)	T1+R1
T2: This magnificent Cathedral	Subject (Participant: Identified) - Unmarked	R2: is the most important church within the worldwide Anglican faith	*T2*+R2
T3: The central 'Bell Harry Tower'	Subject (Participant: Carrier) - Unmarked	R3: dates back to 1498	*T3*+R3
T4: and the crypt	Conjunction + Subject (Participant: Carrier) - Unmarked	R4: to the 11th century	*T4*+R4
T5: It is here	Predicated Theme – Unmarked and Enhanced	R5: [where your guide will tell the story of the shocking murder of Archbishop Thomas Beckett in 1170]	*T5*+R5
T6: and there will also be	Conjunction + Subject &Predicator (Existential 'there') – Unmarked ad Enhanced	R6: time [to explore the narrow streets of Canterbury]	*T6*+R6
T7: There will be	Subject &Predicator (Existential 'there') – Unmarked and Enhanced	R7: an opportunity for lunch in Canterbury	*T7*+R7

The Themes found in this text are mainly unmarked and simple, e.g., 'The central "Bell Harry Tower" dates back to 1498', which indicates an objective message (Francis and Kramer–Dahl, 1991: 354). There is, though, an instance of marked Theme, in which case the dependent clause 'Traveling through Kent' is placed in Theme position, probably with the aim of providing the way to get to the city of Canterbury. The Cathedral's attractions are usually placed in Theme position. When this is not the case, the copywriter seems to use enhanced Themes, i.e., Themes that increase the strength or enhance the prominence of a participant or a circumstantial in the clause complex (Fawcett and Huang, 1995). These Themes counterbalance the objectivity of the unmarked Themes. These are expressed by predicated Themes, e.g., 'It is here where your guide will tell the story of the shocking murder of Archbishop Thomas Beckett in 1170', and existential Themes, e.g., 'there will be an opportunity for lunch in Canterbury'. Their aim seems to be that of emphasising other possibilities apart from visiting a religious landmark (Stoian, 2015), making the visit more interesting.

Regarding the Thematic development in this text, the message does not unfold following any clear pattern, as indicated in Table 2.1. It rather stops at almost every sentence. This is due to the frequent use of new Themes, e.g., 'the crypt' and 'there will be'. They can be, though, retrieved from the previous clauses by pronouns without a clear referent, e.g., 'it', and by synecdoche, e.g., 'the tower'. Their effect can be double-edged, as they can either draw users' attention to that particular item in Theme position or confuse them.

2.3.1.2 The Tower of London

The Tower of London is part of the tour 'Majestic London' together with other famous London's landmarks – Westminster Abbey, Buckingham Palace and St Paul's Cathedral. The text describes the Tower from a historical perspective and points out some of its attractions, such as the Crown Jewels.

The textual analysis (Table 2.2) shows an equal frequency of unmarked and marked Themes. In the case of unmarked Themes, the participants are placed in Theme position, e.g., 'The Beefeaters'. They provide objectivity to the message, objectivity counterbalanced by the subjectivity of markedness. The marked Themes, expressed by preposed attributives, 'Founded by William the Conqueror in 1066-7 and enlarged by successive sovereigns', and circumstances, 'Over the past 1000 years', seem to emphasize the Tower's history and antiquity.

Table 2.2 also shows the lack of multiple Themes and Thematic development. (Please note that *T* in italics indicates new Theme and double square brackets stand for embedded clauses). There are only simple and new Themes. This can indicate a message made up of a series of unconnected sentences in which the copywriter pays more attention to the focalization of information than to the flow of the message.

Table 2.2: The Tower of London: Theme and Thematic structure

Theme	Type of Theme	Rheme	Thematic pattern
T1: Founded by William the Conqueror in 1066-7 and enlarged by successive sovereigns, the Tower of London	Clauses + Subject (Participant: Identified) - Marked & Enhanced	R1: is one of the world's most famous and spectacular fortresses	T1+R1
T2: Over the past 1,000 years	Adjunct (Circumstance: Extent) - Marked	R2: it has been a Royal Palace, an armoury and a place of imprisonment and execution	*T2*+R2
T3: The Beefeaters [who guard the Tower]	Subject (Participant: Actor) – Unmarked + Interpolation	R3: will regale you with stories of its past	*T3*+R3
T4: We	Subject (Participant: Actor) – Unmarked	R4: will visit the Crown Jewels, including the magnificent solid-gold crown [used at the coronation of Queen Elizabeth II], the enormous Cullinan diamonds and the extraordinary Koh-i-Noor	*T4*+R4

2.3.1.3 Edinburgh

The commercial website dedicates an entire tour to the city of Edinburgh, without combining the landmark with other landmarks. As such, the text is slightly longer than the other two analysed so far. It presents a trip by train to the royal city, which can be visited by taking an open-top sightseeing bus. Several attractions are mentioned, e.g., Edinburgh Castle.

The Themes encountered in the text are mainly unmarked and simple. They are expressed by participants like 'the train' and 'you'. There is only one instance of marked Theme, which brings to the readers' attention the place where the open-top sightseeing bus can be taken, i.e., 'right by the station'. As for Thematic development, various patterns are present in equal proportions. The linear pattern is used to unfold the message cumulatively and provide information related to the bus, e.g., 'The hop-on hop-off ticket allows you the freedom [...]. You will see ...'. The continuous pattern maintains the attention on the visitor and the train, e.g., 'You will see [...]. You might want to ...'. Sometimes, new themes, e.g., 'Your tour includes ...', are inserted to emphasize important information, such as the tickets included by the tour indicated in the example. The Thematic structure of the Edinburgh text is presented in Table 2.3. (Please note that *T* in italics indicates new Theme and pointed dashes are used to signal an unclear, but possible Thematic development).

Table 2.3: Edinburgh: Theme and Thematic structure

Theme	Type of Theme	Rheme	Thematic pattern
T1: The train from London's King's Cross station	Subject (Participant- Actor) - Unmarked	R1: *arrives at the Royal City of Edinburgh at around 11.30am*	T1+R1
T2: Right by the station	Adjunct (Circumstance: Location) - Marked	R2: you board the open-top sightseeing *bus* where you'll get a bird's-eye view of this majestic city!	*T2*+R2
T3: The hop-on hop-off ticket	Subject (Participant - Actor) - Unmarked	R3: allows *you* the freedom to leave and rejoin the bus as often as you like	T3+R3
T4: You	Subject (Participant- Senser) - Unmarked	R4: will see the Old Town quarter, the Scotch Whisky Heritage Centre, the gloriously intricate Scott Monument, the Palace of Holyrood House and the brand new Scottish Parliament building	T4+R4
T5: You	Subject (Participant- Senser) - Unmarked	R5: might want to wander through the cobbled city centre streets or stroll along the Princes Street checking out the tartan, whisky and shortbreads that have made Scotland famous	T5+R5
T6: Your tour	Subject (Participant- Identified) - Unmarked	R6: includes tickets to Edinburgh Castle	*T6*+R6
T7: The train back to London	Subject (Participant- Actor) - Unmarked	R7: leaves Edinburgh in the late afternoon returning to London's King's Cross at around 10:30pm	T7+R7

2.3.1.4 Theme and Thematic structure in the British set

Table 2.4 summarizes the Theme and Thematic structure found in the three texts of the British set.

Table 2.4: Theme and Thematic structure in the British set

System	Canterbury Cathedral	The Tower of London	Edinburgh
Theme	+ unmarked	= unmarked and marked	+ unmarked
	+ enhanced	enhanced	-
	+ simple	only simple	only simple
	+ participants as Theme	+ participants as Theme	+ participants as Theme
	- Thematic progression	- Thematic progression	= continuous and linear and new Themes
		all new Themes	

The British texts are rather short and seem to provide an informative and objective message. The Themes are usually unmarked and present the attractions offered by the landmarks promoted. Objectivity is, however, balanced sometimes by the use of enhanced and new Themes. These draw readers' attention to particular attractions. Regarding the Thematic pattern, the texts do not seem to contain a clear Thematic progression. The text on the city of Edinburgh is an exception, thus being continuously centred on the client. Different pieces of information disrupt the flow of the message, which can become, at some point, incohesive and difficult to follow.

2.3.2 Romanian set of texts

The texts included in the Romanian set present the following landmarks: the Monastery of Horezu, the Dacian Fortresses of the Orastie Mountains and Sighisoara. All of them are included in the same tour, 'Grand tour Romania-Unesco tour', which visits the Romanian landmarks that hold the UNESCO designation.

2.3.2.1 The Monastery of Horezu

The Monastery of Horezu appears in the description dedicated to the second day of the tour as an alternative to another landmark, the Church of Saint Nicholas. The text is included in brackets and presents briefly several characteristics of the Monastery. It is very short and contains only one conjoinable clause unit, i.e., one theme.

The only Theme, 'Alternatively you', is unmarked and multiple, as indicated in Table 2.5 (Please note that the underlined words indicate textual Theme). The textual theme expressed by the conjunctive adjunct 'alternatively' highlights the idea of alternative and choice mentioned earlier. Placing the Monastery secondary to other landmark can reduce its importance. As for Thematic progression, there is none, because the text has only one conjoinable clause unit (Table 2.5). The readers' action is focalized in the main clause, while information on the Monastery is added as notes in the dependent clauses. The unfolding of the message seems, thus, unimportant.

Table 2.5: The Monastery of Horezu: Theme and Thematic structure

Theme	Type of Theme	Rheme	Thematic pattern
T1: Alternatively you	Adjunct: conjunctive + Subject (Participant-Actor) - Unmarked	R1: can visit the Horezu Monastery – UNESCO monument founded in 1690 by the Prince Constantin Brancoveanu, considered one of the best examples of 'Brancovean' style, known for the purity of architectural details and richness of details, for its religious composition, ex – voto portraits and ornamental painting	T1+R1

2.3.2.2 The Dacian Fortresses of the Orastie Mountains

The Dacian Fortresses of the Orastie Mountains are presented in the description of the third day of the Tour. Like the Monastery, they are an alternative to the Castle of Hunedoara. The text presents the fortresses' past and their UNESCO recognition.

The Themes encountered in the text and shown in Table 2.6 are unmarked and mainly expressed by participants, e.g., 'The six fortresses form the defensive system of Decebalus'. The frequent presence of multiple Themes, expressed by textual Themes like 'and', provide a cohesive link to the previous messages. Regarding the Thematic development in the text, the message unfolds following two patterns. The Dacian fortresses are picked up from Rheme, showing a linear pattern, and carried on as Theme, by continuous progression ('you can visit *the six Dacian fortresses* [...]. The six fortresses form [...] and *[the six fortresses]* were ...'). The textual analysis of the text is presented in Table 2.6. (Please note that the underlined words indicate textual Theme and the square brackets in italics signal an elided Subject).

Table 2.6: The Dacian Fortresses of the Orastie Mountains: Theme and Thematic structure

Theme	Type of Theme	Rheme	Thematic pattern
T1: Alternatively you	Adjunct: conjunctive + Subject (Participant- Actor) - Unmarked	R1: can visit the *six Dacian Fortresses from the Orastie Mountains* built in the style of murus dacicus, in the period from the first century BC to the first century AD as protection from the conquering Romans	T1+R1 ↓
T2: The six fortresses	Subject (Participant: Identified) - Unmarked	R2: form the defensive system of Decebalus	T2+R2 ↓
T3: and *[the six fortresses]* were declared	Conjunction + Predicator (Process: Causative) - Unmarked	R3: World Heritage by UNESCO	T3+R3

2.3.2.3 Sighisoara

The city of Sighisoara is part of the seventh day of the 'Grand tour Romania-Unesco tour'. Unlike the previous two Romanian landmarks, the city is part of the basic tour, and not an alternative. The text presents several attractions of the city, e.g., Count Dracula. The attractiveness of the city is emphasized by practical information regarding food and accommodation, stimulating, thus, users' interest.

The text contains mainly unmarked and simple themes, which are usually expressed by participants, e.g., 'we'. Even if unmarked, some of the Themes are enhanced, e.g., 'there are several ...', or are expressed by a heavy Subject, e.g., 'Dinner in the house where Count Vlad was born/or an intimate restaurant medieval style and accommodation in Sighisoara in the citadel'. Their aim seems to be drawing users' attention to those particular elements, i.e., the existence of several defensive towers surrounding the city or the possibility to have dinner and accommodation in Sighisoara. The only instance of markedness places the Adjunct 'here' in Theme position ('Here was also born Vlad the Impaler') as if to emphasise that Sighisoara is where Vlad the Impaler was born.

The Themes do not succeed each other in any Thematic pattern of progression (Table 2.7 – Please note that *T* in italics indicates a new theme and double square brackets stand for embedded clauses). The text starts by a linear pattern, which introduces the city, but is continued by a new Theme, e.g., 'there are', which can make the message unclear and thus, confuse readers.

Table 2.7: Sighisoara: Theme and Thematic structure

Theme	Type of Theme	Rheme	Thematic pattern
T1: We	Subject (Participant- Actor) - Unmarked	R1: continue to *Sighisoara*, the 'fortified city' among the very few in Europe, which retains its medieval centre on the top of the hill	T1+R1 ↓
T2: Here	Adjunct (Circumstance: Location) - Marked	R2: was also born Vlad the Impaler, better known as Count Dracula	T2+R2
T3: There are	Subject and Predicator (Existential 'there') - Unmarked and Enhanced	R3: several defensive towers that surround the city named after the trades of their builders: the Tailors' Tower, the Shoemakers' Tower, the Rope Makers' Tower, the Clock Tower, etc.	*T3*+R3
T4: Dinner in the house [where Count Vlad was born]/or an intimate restaurant medieval style and accommodation in Sighisoara in the citadel	Heavy Subject (Participant- Identified) - Unmarked	R4: -	*T4*+R4

2.3.2.4 Theme and Thematic structure in the Romanian set

The Romanian texts display certain similarities (Table 2.8) as far as their Theme and Thematic structure are concerned.

Table 2.8: Theme and Thematic structure in the Romanian set

System	The Monastery of Horezu	The Dacian Fortresses of the Orastie Mountains	Sighisoara
Theme	only unmarked only multiple only participants as Theme - Thematic progression	only unmarked + multiple (textual) + participants as Theme = continuous and linear Thematic progression	+ unmarked only simple + participants as Theme - Thematic progression + new Themes

The message presented by the short Romanian texts appears objective because of the dominance of unmarked Themes. This is sometimes mitigated by the conjunctive adjuncts, which present another voice, that of choice. The lack of any Thematic progression can indicate an unclear or unorganised message, with a split focus between the visitor/tour and the landmark.

2.3.3 Summary of the results

Table 2.9 summarises the results found in the two sets of texts.

Table 2.9: Summary of the results

Theme & Thematic development in the British set	Theme & Thematic development in the Romanian set
+ Unmarked Themes	Almost only unmarked Themes
Only simple Themes	Multiple (textual) Themes
Enhanced Themes	-
No Thematic progression	No Thematic progression
Preference for new Themes	-

The Theme-Rheme analysis of the two sets of texts has indicated certain similarities. Both of them prefer unmarked Themes to present the landmarks and their attractions. These are often not organized in any clear Thematic pattern.

The same analysis has also shown differences in the Theme system of the two sets. The texts dedicated to the British World Heritage Sites combine objectivity with subjectivity in choosing their Themes and Thematic development. The objectivity signalled by the unmarked Themes is often counterbalanced by marked, enhanced and new Themes. The copywriter is the one that chooses which information is more conspicuous and foregrounded. The British message is less compact and cohesive due to its exclusive use of simple Themes and lack of any expressed textual link. In the British set, it seems that the flow of the message is secondary to drawing attention, as it is disrupted frequently by different pieces of information. As for the Romanian set, this is more objective than the British one; there is only one instance of marked Theme. Its message is divided clearly between the visitor/tour and the landmark, each being presented separately. The description of the landmark is subordinated to that of the tour, since it is almost always placed in dependent clauses. The preference for multiple Themes, expressed by textual elements, indicates a connected message. The flow of the message does not seem important, as it has usually no Thematic progression. This lack is not totally unexpected considering the low number of independent conjoinable clause units.

2.4 Conclusion

To conclude, this paper has presented a small-scale study of the Theme-Rheme analysis of two sets of commercial tourist websites belonging to different countries – Great Britain and Romania. The focus has been on three texts, for each set, presenting national World Heritage Sites internationally online. The texts have been first analysed and then compared. Their Theme and Thematic structure have been observed from a systemic functional perspective, following Haliday's model of textual analysis (1985, 1994; Halliday and Matthiessen, 2004) and Daneš' classification of Thematic patterns (1974). The results have shown both similarities and differences between the corpora analysed. Both sets have a preference for unmarked Themes and lack Thematic progression. The British set prefers unmarked Themes, but uses frequently also marked ones. It becomes subjective while drawing attention to certain information by means of marked, enhanced and new Themes. The organization and flow of the message seem secondary to catching users' attention. In the case of the Romanian set, markedness is barely used. The texts present information, linking messages to previous chunks, in an objective manner. The results can be explained by the influence of the context of communication (promotional tourism) and the medium of communication (Internet), which both have an impact on language.

In the end, we would like to stress that Thematic organization can play an important role in the composition of a message as a whole. Our study has shown that different Thematic choices can lead to different messages, some more informative, coherent, cohesive and structured than others. Such details can contribute to the overall decisions related to text production, since the Thematic organization and structure of a text can influence promotion and its perception by users.

References

Calvi, M. (2006). Lengua y Comunicación en el Español del Turismo. Madrid: Arco Libros.

Daneš, F. (1974). Functional Sentence Perspective and the Organization of the Text. In F. Daneš (Ed.), Papers on Functional Sentence Perspective (pp. 106-128). Prague: Academic.

Dejica, D. (2004). On Discourse Structure in Translation: the Concepts of Theme and Rheme. In L. Frenţiu (Ed.), Romanian Journal of English Studies (pp. 47-56). Timişoara: Editura Mirton.

Dejica, D. (2005). Towards a New Model for the Identification of Thematic Information in Discourse. In R Superceanu & D. Dejica (Eds.), Comunicare profesională şi traductologie, (pp. 103-111). Timişoara: Universitatea Politehnica.

Dejica, D. (2009). Identifying and Analysing Theme-Rheme Relations for Discourse Production and Translation. In L. Frenţiu (ed.). Romanian Journal of English Studies, no. 6. Timişoara: Editura Universităţii de Vest, pp. 128-136.

Dejica, D. (2010). Thematic Management and Information Distribution in Translation. Timişoara: Editura Politehnica.

Eggins, S. (1994, 2004). An Introduction to Systemic Functional Linguistics. London: Pinter.

Fawcett, R.P., and Huang, G. (1995). A Functional Analysis of the Enhanced Theme construction in English. In Interface. Journal of Applied Linguistics 10(1) (pp. 113-144).
Francis, G., and Kramer Dahl. A. (1991). From Clinical Report to Clinical Story: Two Ways of Writing about a Medical Case. In E. Ventola (Ed.), Functional and Systemic Linguistics: Approaches and Uses (pp: 339-368). Berlin: Mouton de Gruyter.
Fries, P. (1983). On the Status of Theme in English: Arguments from Discourse. In Micro and Macro Connexity of Text 6(1) (pp. 1-38).
Fries, P. (1995). Themes, Methods of Development, and Texts. In R. Hasan and P. Fries (Eds.), On Subject and Theme (pp. 317-359). Amsterdam: John Benjamins Publishing Company.
Halliday, M.A.K. (1985, 1994). An Introduction to Functional Grammar. London: Hodder Arnold.
Halliday, M.A.K., and C. Matthiessen. (2004). An Introduction to Functional Grammar, 3rd ed. London: Hodder Arnold.
Lombardi Vallauri, E. (1995). A Simple Test for Theme and Rheme in the Clause Complex. In Language Sciences 17(4) (pp 357-378).
Manoliu–Manea, M. (1994). Discourse and Pragmatic Constraints on Grammatical Choices: A Grammar of Surprises. Amsterdam: Elsevier.
McCabe, A. (1999). Theme and Thematic Pattern in Spanish and English History Texts. PhD. diss., Aston University.
Morgan, N., and Pritchard, A. (2000). Advertising in Tourism and Leisure. Oxford: Butterworth-Heinemann.
Stoian, C.E. (2015). The Discourse of Tourism and National Heritage. Newcastle upon Tyne: Cambridge Scholars Publishing.
Thompson, G. (1996). Introducing Functional Grammar. London: Arnold.
Whittaker, R. (1995a). Theme, Processes and the Realization of Meanings in Academic Articles. In M. Ghadessy (ed.) Thematic Development in English Texts (pp. 105-128). London: Pinter.
Whittaker, R. (1995b). More than Mode: Writer's Purpose and Choices at Theme. In I. Vázquez and A. Harnero (eds.) Current Issues in Genre Theory (pp. 83-130). Zaragoza: Mira Editores.

Simona Şimon

3 Necessary and Luxury English Loanwords in Some Romanian Online Newspapers and Magazines

3.1 Introduction

After the 1989 fall of the communist regime, but particularly after the 2007 accession to the European Union, Romania has kept on strengthening the socio-economic, cultural, religious, and political ties with the rest of the world. In such a situation, 'a "contact language" between persons who share neither a common native tongue nor a common (national) culture' (Firth, 1996: 240) was obviously needed. As English is the 21st century 'contact language' (Crystal, 2012), the Romanian government has encouraged the teaching of English in the Romanian public schools. English has, thus, become the most taught foreign language in the Romanian public schools as the 2012's 'Key Data on Teaching Languages at Schools in Europe' published by The Education, Audiovisual and Culture Executive Agency of The European Commission proves it (http://eacea.ec.europa.eu/education/eurydice/documents/key_data_series/ 143EN.pdf).

A direct and expected consequence of such a situation is the increasing number of borrowed English words in the Romanian language. Besides the socio-economic, cultural, religious, and political relations among various nations, other factors that contribute to the borrowing of foreign words into a language in general, and in Romanian in particular, may be worth mentioning. For example, the continuous scientific and technical development that characterises the contemporary society favours the borrowing of foreign words (Hristea, 1984: 39-40). The scientific discoveries and the fast technical progress bring about new concepts which are quickly adopted by the Romanian language. One of the reasons for which such an adoption rapidly takes place is the fact that the new inventions are swiftly replaced or complemented by others, so that there seems to be no time to find Romanian equivalents for the new concepts that keep on appearing one after another. The scientific and technical development contribute to the improvement of people's lifestyle, and thus, to the change in their perspective on life, which is viewed as another factor that creates the necessary framework within which the borrowing of foreign words into Romanian may take place (Hristea, 1984: 39-40). The borrowing process is also favoured by the way in which the donor culture, and implicitly language, is perceived by the recipient culture and language (Haspelmath, 2009: 35). The superior, more advanced culture tends to be valued by the inferior, less developed culture. In our case, the English-speaking cultures are considered by the Romanians to be role models, and as such, English is more prestigious than Romanian. In such a context, the use of English words when speaking or writing

Romanian, although it is not always necessary, confers the stretch of discourse, and in some cases, the sender, a certain degree of prestige. Finally, the grammar of the two languages may also favour the borrowing of words from one language into the other (Haspelmath, 2009: 35). For example, from a grammatical point of view, nouns 'receive, not assign, thematic roles' (Myers-Scotton, 2002: 239), and therefore, they tend to be less adapted than verbs and easier to borrow (Haspelmath, 2009: 35; Șimon & Suciu, 2014).

Taking into account the functional need of the Romanian language to borrow English words, the present study focuses on the English loanwords present in thirty Romanian online newspapers and magazines, which were analysed in September 2014, and on their classification into necessary and luxury loanwords. In the final part of the present study, some concluding remarks with respect to the most frequently encountered English loanwords are put forward.

3.2 Theoretical background

The borrowing of foreign words into a language has concerned many linguists interested in the evolution of various languages. For example, in Romania, studies were made on the origin of the loanwords present in the Romanian language (Hristea, 1984), on the phonetical, morphological, lexical, and semantic adaptation of the foreign words to the Romanian language (Avram, 1997; Ciobanu, 1991; Pârlog, 1971; Stoichiţoiu-Ichim, 2006), or on the field of activity from which they have been borrowed (Greavu, 2011; Stoichiţoiu-Ichim, 2006). The aforementioned studies (Avram, 1997; Ciobanu, 1991; Greavu, 2011; Hristea, 1984; Pârlog, 1971; Stoichiţoiu-Ichim, 2006) also try to clarify the terminological issues regarding the words borrowed from one language into the other, making recourse to both the national and international literature. In the following sections, only some of the most important and relevant aspects will be discussed.

The terms 'loanwords' or 'borrowings' designate the 'words taken from one language and transferred into another' (Șimon & Suciu, 2014: 6). The two languages involved in this process are the donor language and the recipient language. The donor language is also called the source or original language, while the recipient language is termed the target or replica language as well (Fasold & Connor–Linton, 2006: 294; Furiassi, Pulcini & Gonzales, 2012: 12). The first one lends words to the second one, but these are never returned: 'When one language takes lexemes from another, the new items are usually called *loanwords* or *borrowings* – though neither term is really appropriate, as the receiving language does not give them back.' (Crystal, 2005: 126)

Researchers on this topic have noticed that the loanwords are taken differently into the recipient language. So, Haugen (1950: 210-231) distinguishes between loanwords, loanblends and loanshifts. Loanwords preserve their form and meaning in both languages. For example, the word *airbag* is an English loanword in Romanian, as it is used with the same spelling and meaning both in the donor language and in the recipient language. Loanblends consist of two parts, one borrowed from the donor language and one belonging to the recipient language. For instance, the Romanian word *scanare* is a loanblend in Romanian because it comes from the English *scanning*. In this particular example, the English suffix *–ing* is replaced by the Romanian suffix *–are*. Loanshifts copy only the meaning of the word taken from the donor language. A common example of loanshift is the Romanian word *zgârie-nor* which actually is a translation of the two words that compound the English word *skyscraper*. Both the Romanian *zgârie-nor*, and the English skyscraper designate the same object.

Apart from Haugen's classification of loanwords, which mainly takes into account their form, another important classification is made from the perspective of their contribution to the enrichment of the recipient language. For example, Myers–Scotton (2002: 239) distinguishes between cultural and core borrowings. The cultural borrowings designate new objects or concepts, for example, *espresso*, *banner*, *Zeitgeist* or *Apartheid*, while the core borrowings duplicate lexemes already existing in the recipient language, for example, *popcorn*, *party*, or *look*. Other authors label the presence of such foreign words in a recipient language as necessary and luxury loans (Danesi & Rocci, 2009: 162; Furiassi, Pulcini, & Gonzales, 2012: 46). The necessary loans are foreign words borrowed by the recipient language because they designate new objects, institutions, concepts or situations that have not been termed before. The luxury loans, as they are called, duplicate the already existing words in the recipient language, although they do not add any new senses. In other words, some scholars prefer the dichotomy cultural - core borrowings, while others the dichotomy necessary – luxury loanwords. Both of them actually designate the same linguistic phenomena.

Taking into account the contribution of the English loanwords to the enrichment of the Romanian language, some linguists have expressed their opinion on this matter. For example, Ciobanu (1997) considers that the necessary loanwords are adopted 'for their informative function' (1997: 91), as they describe new realities, while the luxury loanwords are adopted because they confer the user 'cultural and social prestige, or, to put it more bluntly, there is a good deal of snobbishness involved in such preferences' (Ciobanu, 1997: 92). Avram (1997: 9) and David and Tălmăcian (2013) have a more tolerant attitude towards the borrowing of the English words into Romanian. They consider that English 'has an important Latin component' (David & Tălmăcian, 2013: 41) and as such, the English loanwords, even the luxury ones 'continue the process of re-Latinization of the Romanian language' (2013: 41). This means that the new influx of English loanwords into Romanian only revives the language without putting its identity in danger (Sim, 2007: 988, cited in David & Tălmăcian, 2013: 41).

3.3 Necessary and luxury English loanwords in some Romanian online newspapers and magazines

Most of the corpus on which the present study has been carried out was also used in a previous article which mainly dealt with the most frequently encountered speech category to which the loanword belonged (Şimon & Suciu, 2014). The corpus is made up of 205 loanwords (28 adjectives / adverbs, 170 nouns / noun phrases and 7 verbs / phrasal verbs) identified in thirty Romanian online newspapers and magazines in September 2014. For the completion of the present study some of the latest dictionaries dealing with the loanwords in Romanian (Bucă, 2014; Ciobanu, 1996; Crăcea, 2012; Dejica, 2013), but also some other traditional or online dictionaries (Bantaş, 1999; https://dexonline.ro/; http://ro.bab.la/; http://www.oxforddictionaries.com/) have been checked.

The adjectives and adverbs identified in some Romanian online newspapers and magazines are: casual, chic, classic, classy, collected, cool, easy, fit, funky, fresh, fulltime, glam, live (music), nude, offline, online, open, second-hand, sexy, skinny, single, slim, smart, smokey, stylish, sun-kissed, trendy, and wireless. The words *live, offline, online* and *wireless* are the only ones that do not have a Romanian corresponding word, which means that they are the only necessary loanwords in the above list of adjectives / adverbs, all the others being the luxury ones.

The English nouns and noun phrases that were spotted in the selected online newspapers and magazines are alphabetically listed in Table 3.1.

In the table below, the words *airbag, babysitting, banner, blog, blogger, brainstorm(ing), burger, casting, computer, derby, desktop, drive-in cinema, email, Facebook, Facebook-feeds, hardware, hippie, hipster, jeans, laptop, marketing, modelling, site, SMS, snowmobil, social media, software, spam, start-up, voucher, web,* and *web developer* do not have a Romanian corresponding word, and therefore, are considered necessary loanwords as they enrich the existing Romanian vocabulary.

Finally, a few verbs and phrasal verbs were singled out: to feature, to leather up, to look, to rebrand, to plug in, to drag and to drop. From the listed verbs and phrasal verbs, there are only two that do not have a Romanian equivalent, namely *to leather up* and *to rebrand*, which are, thus, considered to be necessary loanwords.

The present study reveals that from the twenty-eight adjectives and adverbs, four are actually necessary loanwords, the other twenty-four being luxury loanwords with no informative function. Thirty-two of the one hundred and seventy nouns and noun phrases are necessary loanwords. Out of the seven verbs / phrasal verbs, there is one verb and one phrasal verb that can be considered necessary loanwords, the remaining five verbs and phrasal verbs have no informative function at all and are, therefore, considered to be luxury loanwords. In other words, almost 14% of the adjectives / adverbs, 19% of the nouns / noun phrases, and 29% of the verbs / phrasal verbs are necessary loanwords. The other 86% of the adjectives / adverbs, 81% of the nouns / noun phrases, and 71% of the verbs / phrasal verbs are luxury loanwords. From the

Table 3.1: English nouns and noun phrases used in some Romanian online newspapers and magazines

airbag	device	jeans	play list	social media
aqua park	discount	job	pocket-size	software
baby	display	jogging	pony tail	song writer
babysitting	dress	kit	pool	soul
backstage	drive-in cinema	hair-stylist	powder	spam
banner	duty-free shop	hit	quiz	speaker
beauty	DVD	hotline	rating	star
blush	email	label	remix	start-up
body	entertainment	laptop	review	steward
blog	eye-liner	laundry service	rock	story
blogger	Facebook	life	room-service	streaming
boom	Facebook-feeds	like	statement	tag
brainstorm(ing)	fashion	link	stick	target
boyfriend	fitness	(live) music	supermarket	team-building
browser	freestyle	look	sailing	tip
burger	fun	lounge	science	track
business	fun fact	love	screening	travel
cardigan	gadget	make-up	scrub	trench
career	glamour	marketing	scuba diving	trend
casting	grill	mascara	selfie	trick
catwalk	hair	modeling	server	T-shirt
CD	hardware	mountain bike	Service charge	TV
click	high definition	multiplayer	sex-appeal	twist
club	high performance	must-have	shooter	update
college	high-tech	nails	shopping	upgrade
cloud	hippie	news	showbiz	vintage
comment	hipster	newsletter	single	voucher
computer	hobby	office	site	web
cover	home	(open) bar	skill	webcam
cruise	icon	outfit	smart phone	web developer
cycling	inbox	party	SMS	weekend
derby	internship	party girl	snake print	window
designer	infinity	PC	snowboard	work
desktop	jazz	piece	snowmobil	yacht

total number of two hundred and five loanwords identified in thirty Romanian online newspapers and magazines in September 2014, thirty-eight are necessary loanwords and one hundred and sixty-seven are luxury loanwords. To put it differently, almost 19% are necessary loanwords, while 81% are luxury ones. This rapport, which shows a tendency of the Romanian online journalists to favour the use of luxury loanwords of English origin, is clearly illustrated in the figure below.

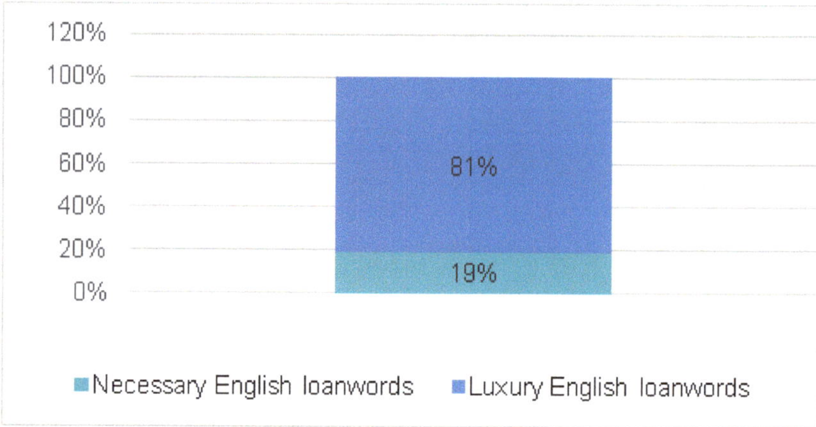

Figure 3.1: English necessary and luxury loanwords in some Romanian online newspapers and magazines

3.4 Conclusions

The present study described the context in which English became the most often taught foreign language in the Romanian public schools, and therefore, adopted by many Romanians for their private and professional communication. Apart from this, various other factors that might contribute to the borrowing of foreign words in the recipient language were identified. The loanwords, loanblends and loanshifts were then defined in order to clarify the theoretical concepts that were useful for the present study. The loanwords were further classified into necessary and luxury loanwords.

The corpus on which this analysis was carried out was made of two hundred and five English loanwords that were singled out in September 2014 in thirty Romanian online newspapers and magazines. The identified English loanwords were twenty-eight adjectives / adverbs, one hundred and seventy nouns / noun phrases, and seven verbs / phrasal verbs. The analysis outlined that 14% of the adjectives / adverbs, 19% of the nouns / noun phrases, and 29% of the verbs / phrasal verbs were necessary loanwords. That meant that the other 86% of the adjectives / adverbs, 81% of the nouns / noun phrases, and 71% of the verbs / phrasal verbs were luxury loanwords. To put it differently, from the total number of two hundred and five English loanwords, thirty-eight were necessary loanwords and one hundred and sixty-seven were luxury loanwords. It followed that almost 19% of the selected English loanwords present in some Romanian online newspapers and magazines were necessary loanwords and 81% were luxury ones.

The present study showed a huge tendency of the Romanian online journalists to prefer English luxury loanwords in order to describe certain concepts, objects, in a word realities. As pointed earlier, the linguistic choices reflect the author's perception on the relationship between the donor and the recipient language, that is the prestige of the donor language and culture in comparison to that of the recipient language and culture on one hand and on the other, they mirror the author's wish to stand out, to attract the others' attention even if this means to be labelled as a snob.

References

Avram, M. (1997). Anglicismele în limba romană. Bucureşti: Editura Academiei Române.
Bantaş, A. (1999). Dicţionar englez-român, Bucureşti: Teora.
Bucă, M. (2014). Dicţionar de neologisme, Bucureşti: Meteor Press.
Ciobanu, G. (1991). The Phonological Adaptation of Anglicism in Romanian. In R. Filipović (ed.), The English Element in European Languages, vol. 3 (pp. 30-59). Zagreb: University of Zagreb.
Ciobanu, G. (1996). Anglicisme în limba română, Amphora: Timişoara.
Ciobanu, G. (1997). Romanian Words of English Origin, Timişoara: Amphora.
Crăcea, E. (2012). Dicţionar de neologisme, Constanţa: Steaua Nordului.
Crystal, D. (2005). The Cambridge Encyclopedia of the English Language, Cambridge: Cambridge University Press.
Crystal, D. (2012). English as a Global Language, Cambridge: Cambridge University Press.
Danesi, M., & Rocci, A. (2009). Global Linguistics: An Introduction, Berlin: Mouton de Gruyter.
David, I., & Tălmăcian, E. (2013). Linguistic and Cultural Dimensions of Lexical Borrowings from English into Romanian, Synergy, Volume 9, Issue 1 / 2013, 38-45.
Dejica, D. (2013). Dicţionar englez - român, român - englez cu termeni şi expresii utilizate in media digitală, Politehnica: Timişoara.
Fasold, R.W., & Connor–Linton, J. (eds.). (2006). An Introduction to Language and Linguistics, Cambridge: Cambridge University Press.
Filipović, R. (ed.). (1991). The English Element in European Languages, Vol. 3. Zagreb: University of Zagreb.
Firth, A. (1996). The discursive accomplishment of normality. On "lingua franca" English and conversation analysis, Journal of Pragmatics, Volume 26, Issue 2, Elsevier, 237-259, http://www.sciencedirect.com/science/article/pii/0378216696000148 (accessed 27.10.2014).
Furiassi, C., Pulcini, V., & Gonzales, F.R. (eds.). (2012). The Anglicization of European Lexis, Amsterdam, Philadelphia: John Benjamins Publishing Company.
Greavu, A. (2011). A Corpus-Based Approach to Anglicism in the Romanian Economic Press, Studies in Business and Economics, Vol. 1, Issue 6 (2), 37-45.
Haspelmath, M. (2009). Lexical borrowing: Concepts and issues. In M. Haspelmath, & U. Tadmor (eds.), Loanwords in the World's Languages: A Comparative Handbook (pp. 35-54). Berlin: De Gruyter Mouton.
Haugen, E. (1950). The Analysis of Linguistic Borrowings, Language, Vol. 26, No. 2, 210-231, http://dx.doi.org/10.2307/410058 (accessed 01.11.2014).
Hristea, Th. (coord.). (1984). Sinteze de limba română, Bucureşti: Albatros.
Myers-Scotton, C. (2002). Language contact: Bilingual encounters and grammatical outcomes, Oxford: Oxford University Press.
Pârlog, H. (1971). Termeni de origine engleză în publicistica română contemporană, Analele Universităţii din Timişoara IX, 55-59. Timişoara: Tipografia Universităţii.

Sim, M. A. (2007). English borrowings – means of re-Latinisation of the Romanian language, http://steconomice.uoradea.ro/anale/volume/2007/v2-the-impact-of-foreign-languages- on-the-development-of-the-business-environment/11.pdf (accessed 7.03.2013).
Stoichiţoiu-Ichim, A. (2006). Aspecte ale influenţei engleze în româna actuală. Bucureşti: Editura Universităţii din Bucureşti.
Şimon, S., & Suciu, L. (2014). English Loanwords in Some Romanian Online Newspapers and Magazines, Scientific Bulletin of the Politehnica University of Timişoara - Transactions on Modern Languages, Vol. 13, Issue 1/2014, 05-10. Timişoara: Editura Politehnica (available also online at http://www.cls.upt.ro/files/cls/Publicatii/Buletinul%20Stiintific/Vol.%2013,%20 2014/01%20Simon_Suciu.pdf).
*** (2012), Key Data on Teaching Languages at Schools in Europe, http://eacea.ec.europa.eu/education/eurydice/documents/key_data_series/143EN.pdf, The Education, Audiovisual and Culture Executive Agency, 2012, 60-64 (accessed 27.10.2014).
***, https://dexonline.ro/
***, http://ro.bab.la/
***, http://www.oxforddictionaries.com/

Appendix: Romanian Online Newspapers and Magazine

Adevărul.ro, http://adevarul.ro/(accessed September 2014)
Bravo, http://bravonet.ro/(accessed September 2014)
Click!, http://www.click.ro/(accessed September 2014)
Connect, http://www.connect.ro/(accessed September 2014)
Cosmopolitan, http://www.cosmopolitan.ro/(accessed September 2014)
Diva Hair, http://www.divahair.ro/(accessed September 2014)
Fashion Days, https://www.fashiondays.ro/campaigns/women/ (accessed September 2014)
Femeia., http://www.femeia.ro/(accessed September 2014)
Glamour, http://www.glamour.ro/(accessed September 2014)
Historia.ro, http://www.historia.ro/(accessed September 2014)
HotNews.ro, www.hotnews.ro(accessed September 2014)
Jurnalul.ro, http://jurnalul.ro/(accessed September 2014)
Kudika, http://www.kudika.ro(accessed September 2014)
Libertatea, http://www.libertatea.ro/(accessed September 2014)
Live News, http://www.livenews.ro/(accessed September 2014)
Network World Romania, http://www.networkworld.ro/(accessed September 2014)
One.ro, http://www.one.ro/(accessed September 2014)
PC Games 4 Fun, http://www.pcgames.ro/(accessed September 2014)
PC World, http://pcworld.ro/(accessed September 2014)
ProMotor.ro, http://www.promotor.ro/(accessed September 2014)
Psychologies, www.psychologies.ro(accessed September 2014)
Romania Lifestyle, http://romanialifestyle.ro/(accessed September 2014)
Reader's Digest Romania, http://www.erd.ro/(accessed September 2014)
Smart News, http://www.smartnews.ro/(accessed September 2014)
Spy News, http://spynews.ro/(accessed September 2014)
Terra Magazin, http://www.terramagazin.ro/(accessed September 2014)
Top Gear, http://www.topgear.ro/(accessed September 2014)
Unica, http://www.unica.ro/(accessed September 2014)
Viva!, http://www.viva.ro/(accessed September 2014)
Ziarul financiar, http://www.zf.ro/(accessed September 2014)

Diana Oţăt

4 Corpus Linguistics Outcomes and Applications in the Digital Era

4.1 Introduction

Prominent scholars highlight the key role of word use in the development of corpus linguistics as the study of linguistic phenomena by means of extensive collections of machine-readable texts, i.e., by means of corpora. The steady evolution of corpus linguistics has been primarily motivated by the linguists' need to understand how words are actually used in natural languages, which most common words tend to be used in certain contexts, what is common and what is uncommon for certain language variations (including specialisms), thus leading to the first outcomes provided by corpus-based approaches, i.e., words lists and synonymous terms. Hence, mainstream literature pinpoints the emergence of corpus-based investigations as early as 1755, when, as endorsed by Biber *et al.* (2006: 22), Johnson used a corpus of texts to gather authentic uses of words that he then included as examples in his dictionary – a first step made towards the understanding of the patterns of use associated with a word.

A less popular branch of linguistics, though widely explored in the 1940s and 1950s, *statistical linguistics*, has also contributed to the development of what nowadays corpus linguistics has become through innovative mathematical theories of information. Yet, due to the lack of computer-assisted processing tools back at that time, it proved to lack productivity and effectiveness.

Another branch of linguistics closely related to the current state-of the-art of corpus linguistics was what 'older linguists, of the heyday in the 1950s', such as Harris, Fries or Hill and other American structuralists regarded as 'descriptive linguistics' (Leech, 1992: 106), i.e., the scholars' aim at describing the corpus under investigation. Accordingly, endorsing the flexible typology of *descriptive linguistics* towards theory construction, Leech (*ibidem*) highlights the less abstract nature of its outcomes, particular to one language, where linguistic phenomena are more easily to localise, observe and analyse. In this context, we grow aware of one main characteristic of corpus linguistics, namely that most corpus-based analyses are applied to data inquiries specific to individual languages.

In the 1980s, linguists registered a rebirth of corpus linguistics, which by then had already indicated a close connection with *quantitative linguistics*, a specialised research branch that promotes the need for quantitative methods in language study, which, according to various linguists, are also frequently used in most other disciplines as they can provide reliable outcomes when it comes to the description of language in terms of frequency and infrequency rates.

However, this diachronic examination of the development of corpus linguistics differs considerably from the status it has acquired in the digital era. For contemporary researchers, corpus-based analysis does not only serve the mere purpose of dictionary making. Theorists and practitioners alike go far beyond the normative function of corpus-based analyses, concerned with simple inventories of linguistic structures, charting new territories in the development of corpus linguistics oriented towards qualitative and functional interpretations of quantitative research. It represents, as postulated by Svartvik (1992: 8), a way 'to take a look at real manifestation of language when discussing linguistic problems', for as 'corpus linguistics is not the heaping of data for its own sake, but rather the investigation of data for scientific purposes' (*ibidem*).

4.2 Corpus Linguistics in the Digitalised Era

It is common knowledge that the dominant role in corpus linguistics is played by modern technology. As previously mentioned, corpus linguistics has been regarded as an operational framework in language study, rather than an isolated domain of study. It is not a monolithic system providing fixed sets of homogenous methods and procedures applied to language investigation, but, as defined by McEnery and Hardie (2012: 5), 'a heterogeneous field'. Applied to the study of the language, corpus-based analysis aims at investigating particular linguistic structures, the way they occur in different contexts and the functions they acquire. The current perspectives in corpus linguistics indicate the extensive use of machine-readable texts as the appropriate resources, the raw material on which to study specific linguistic issues and phenomena.

As put forward by scholars, such as Kennedy (1998) or Biber et al. (2006), computer technology advances not only have multiplied the investigation perspectives where to apply corpus-based analysis, but have also provided novel benefits and backups in comparison to, previous research studies. Undoubtedly, the first significant advance that upgraded this hybridised field is the storage capacity of very large databases of natural language that can be compiled from a wide range of sources. Today, we can save, store and organise on our own computers ample writings or large chunks of texts, being, thus, able to carry out thorough linguistic analyses that are no more limited to sentence-length excerpts.

Modern software and computer-assisted tools secure comprehensive linguistic analyses that are more accurate and reliable. In this climate of opinion, Biber *et al.* (2006) also state that 'unlike human readers that are likely to miss certain occurrences of a word, computers can find all the instances of a word in a corpus and generate an exhaustive list of them'. However, it is worth mentioning that corpus linguistics is not only concerned with words frequency, for as other branches of linguistics, such as phonetics, syntax and any another aspect of linguistics may be investigated via corpora analysis while applying and combining a series of corpus-based investigation methods. In line with the prominent figures in the field of linguistics, we can endorse that computer-assisted analysis methods can be applied to various types of corpora, aiming at investigating particular patterns of word associations on a far more complex model than it is possible manually.

Among the core features of computer-assisted corpora analysis applied to language study, we can mention the empirical character of corpus-based analysis, i.e., the investigation of actual language structures in authentic texts, namely in corpora, which represent the basis for the analysis designed and implemented. As previously indicated, dedicated software and computer-based analysis tools have been extensively applied within this field of research, which led to the implementation and operation of hybridised techniques, which rely heavily on quantitative and qualitative methods alike. As endorsed by Biber *et al.*, computer-assisted investigation has provided not only reliable analysis methods, but also consistent outcomes, enabling individuals actively interact in validating complex linguistic findings, while 'the computer takes care of record-keeping' (*ibid*: 6).

Two main corpus-based research directions have been indicated by the authors mentioned above, where linguistic features are not investigated as isolated occurrences, but as systematic associations with other features. Accordingly, we distinguish between 'linguistic associations of the features' that encompass lexical and grammatical associations and 'non-linguistic associations of the features', i.e., distribution across registers, dialects and time periods (*ibidem*).

Concerned with the study of linguistic features by means of corpus-based analyses, it is worth mentioning that corpus-based research can be applied to grammar on the word level as well as on the sentence and discourse levels. Also, register and text typology can be investigated by means of corpora study.

As far as grammar study is concerned, corpus-based analysis has recently registered fruitful advances towards novel interdisciplinary areas of investigation. Thus, even though former linguists highlighted the descriptive approach applied to the grammar study, prescriptive methods tend to be used more frequently, for example, in order to establish the variables framing the syntax of language. By applying corpus-based analysis, new research perspectives envisaged the patterned use of grammatical features in texts or the variation of language in use. The investigations of natural language texts have enabled linguists to register and further investigate the routinised ways that individuals tend to use the grammatical resources of a language. Linguists postulated that, by investigating the distribution of various structures, the association patterns between grammatical structures and other linguistic and non-linguistic factors that influence the individuals' linguistic choices, all of them recorded in authentic texts, considerable outcomes and research advances have been achieved. Such fruitful headways were not only reached at the grammatical level, but also in terms of language evolution. Its dynamics and the implementation of the most appropriate strategies to bridge socio-cultural and linguistic differences is an enduring benefit for closely related fields of research, such as translation studies, sociolinguistics or cultural studies.

4.3 Corpora Design and Compilation in the Digital Era

The central aim of the present paper is not only to highlight the considerably value-added and reliable outcomes provided by computer-assisted corpus-based analysis, but also to pinpoint the highly effective dedicated software and computer-assisted tools applied in corpora design and compilation.

We should first mention that, paradoxically, the concept of *corpus* was initially used to designate various non-linguistic collections, such as the *Corpus Juris Civils* introduced by the emperor Justinian in the sixth century and regarded by numerous scholars as a compilation of early Roman law and legal principles, which also illustrated particular cases and provided clarifications of new laws and future legislation to be put in effect (see Jan Svartvik, 1992). However, linguistics corpora are regarded as collections of texts 'assumed to be representative of a given language, dialect, or other subset of a language, to be used for linguistic analysis', as defined by Francis (1992: 17). Thus, in language study, corpora have been primarily used for linguistic analysis, a feature that according to Francis (*ibid*:19), differentiates them from other types of corpora or large text collections, such as anthologies, for example, the *Oxford Book of English Verse*, whose purpose is literary.

Other examples of corpora types are lexicographical corpora, compiled in the process of making dictionaries, for example, the *Oxford English Dictionary*, edited in the 19th century by Murray or the *Merriam-Webster Dictionary* edited in the 20th; dialectological corpora compiled for the purpose of designing dialect atlases, for example, in the *Middle English Dialect Atlas*, issued in 1981 by Benskin, or grammatical corpora, among which Quirk's *Survey of English Usage* published in the 20th century is the most well known.

Starting with the early 1960s, modern concepts of corpora generally indicate the use of large collections of texts available in machine-readable forms, of which the most representative one is the *Brown University Standard Corpus of Present-Day American English* compiled by Henry Kucera and W. Nelson Francis and which remains a sample of present-day English for use with digital computers. In the next two decades, it was followed by the creation of the *British National Corpus* (BNC), started in 1991.

Given the broad and multi-faceted research directions that emerged in the field of linguistics, as well as in previously mentioned close related domains, where corpora investigations are applied in order to achieve novel quantitative and qualitative outcomes, McEnery and Hardie (2012: 12) put forward a series of criteria applied to distinguish between the types of corpus-based investigations:

- mode of communication
- corpus-based *versus* corpus-driven communication
- data collection regime
- the use of annotated *versus* unannotated corpora
- total accountability *versus* data selection
- multilingual *versus* monolingual corpora

By establishing these criteria, the authors aimed at featuring a typology of corpus linguistic research framed by the principles of corpora use.

Albeit corpus-based analyses have by no means been restricted to the English language, it is common knowledge that corpora investigations applied to the study of the English language have provided the most relevant and significant language study perspectives in corpus linguistics during the past period, leading to a boost and proliferation in English studies generally. Among the most frequently approached research perspectives carried out by means of corpus analysis, we could mention the study of language variation, dialect, register and style, where authentic samples of different areas of language use have been compiled and investigated in order to validate the diverse range of language users or the close analysis of frequency rates of particular linguistic structures in different language varieties or in certain specialised languages.

Thus, corpora are far from being considered mere language samples aimed to provide useful illustrative examples, but genuine theoretical resources, which, implemented in a series of applied fields of research, such as language teaching, translation studies or even machine translation and dedicated processing software (spelling, grammar and style), have provided essential sources of information leading towards generalisations about the language and language use. Furthermore, online freely available corpora are meant to ensure linguists from all over the world a user-friendly access to language materials that would otherwise be difficult or impossible to obtain. Conversely, linguists who are non-native speakers can use such corpora for further research and practice.

Having established that corpora design and compilation in the context of new cutting-edge technologies is a key concern of the present paper; it is also noteworthy mentioning that corpora fall into two major categories, i.e., *general corpora* – designed to investigate a given language as a whole – and *specialised corpora* – designed to answer more specific research questions, mainly used in the study of special languages or closely connected applied fields of research. Accordingly, while general corpora are thoroughly organised in order to have a long 'shelf-life', as claimed by Lüdeling and Kytö (2008: 154), specialised corpora aimed for the study of certain linguistic items in certain contexts or circumstances are much more rapidly constructed. The authors consider that most of the corpora can be understood as a collection of sub-corpora which display relative homogenous features. Inevitably the current corpora design is based on 'a template of variables that creates a number of cells, each of which constitutes a sub-corpus' (*ibidem*). If a linguist aims at investigating the specialised domain of the written legal language, s/he may start designing her/his corpus by collecting, storing and organising various types of legal documents, such as contracts, laws, treaties, etc. that may constitute the cells or the sub-corpora.

Aston and Burnard (1998: 29) put forward a series of references applied in corpora design, i.e., *field*, *tenor* and *mode* variables, which, for example, can be easily recognisable amongst the most representative criteria of the British National Corpus

(BNC) that includes *field* (the subject-matter of written texts), *tenor* (spoken texts), and *mode* (books, periodicals).

A central issue in corpora design and particularly in the design of specialised corpora is how to establish the most appropriate variables and how the dynamics of the sub-corpora should function, considering the fact that most of the times such specialised corpora reveal a multi-dimensional character. Nevertheless, as argued by Lüdeling and Kytö (2008), in most situations it is obviously not possible to collect all the texts that constitute the research object into a corpus. In this respect, the authors advocate that the selection of a subset of the texts aimed for analysis stands as a reliable method. Moreover, the scholars recommend that establishing of strict sampling techniques may lead to reliable and operative corpus design. We would tend to emphasise that the contents of a corpus designed for research purposes need to be carefully considered, though some will argue that the design of the corpus will depend more on what is freely available in an easily converted format than on other criteria, thus pinpointing towards the benefits provided by the digital era.

Within this context, a series of other very important aspects that need to be considered in the corpora design can be mentioned, i.e., software limitations, copyright, and text availability.

It is a truism that dedicated software provides an overwhelming capacity to store databases under different shapes (written, spoken, images, graphics, etc.); various linguists have pointed out that readily available software packages, such as *WordSmith Tools* (Scott, 2004), can deal with a corpus of tens of millions of tokens in size. Lüdeling and Kytö (2008: 157) also mention the existence of larger corpora that often work with software, which demands that the tokens are converted into digits, which would facilitate the software to process more quickly. Be that as it may, specialists from linguistics and IT alike advocate that complex operations on corpora that may involve hundreds of millions of words can take some time to complete.

Furthermore, another aspect that needs to be taken into consideration is that the storage and further use of the electronic version of published texts remains a crime in many of the EU Member States and worldwide as well, if no copyright permissions are legalised. Thus, due to the fact that such permissions are often difficult to gain, the availability of some corpora may be restricted.

Another aspect highlighted also by Lüdeling and Kytö (*ibid*: 158) refers to the availability of texts. In this respect, the authors mention the design limitations imposed by the corpora designers who wish to include in their corpora 'spoken texts from an era before the invention of tape recorders'. The accuracy of the data as a representation of the actual speech is always questionable. Even though written texts are easier to obtain, there are some situations when extensive written texts are difficult to be incorporated into a corpus, unless very large storage facilities are available for scanning and keying.

4.4 Application: A Model for Computer-Assisted Corpus Design and Analysis

In what follows, we aim at exemplifying a computer-assisted model for corpus design and corpus analysis that can be applied to specialised corpora used for the investigation of specialised languages as well as to other fields of research, such as language teaching or translation training programmes.

The model proposed is based on the MAXQDA 11 for Windows professional software dedicated to qualitative and mixed methods data analysis. As highlighted by its promotors, MAXQDA provides a variety of research methods and approaches. By means of this computer-assisted tool, we can organise, encode, annotate, and interpret an array of data. Moreover, the analysis outcomes can be generated in easy-to-read reports, visualisations, Excel sheets, while enabling the researchers to work interactively and share the results among each other.

Applying MAXQDA to the study of language, we simulated first the possibilities of the corpus design. Providing the fact that MAXQDA allows the import of various format types, a possible corpus for language analysis may encompass a wide range of written and spoken texts, as well as images, graphics or tables in TXT, RTF, DOC/X, PDF, JPG, GIF, TIF, and PNG format. Media files can be also selected here, enabling the designer(s) to set up even a multimodal corpus.

The software allows the users to design their corpus in a project-based format, where each group member can actively participate, save memos and observations, and apply them both to the design and the analysis of the corpus.

Moreover, the software allows the users design various corpora simultaneously, which can be further structured in a sub-corpora, if organised in separate document groups.

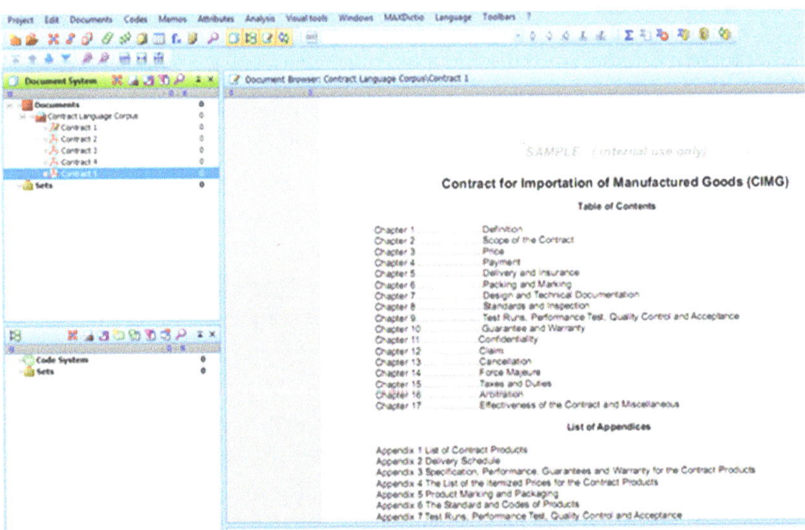

Figure 4.1: Print Screen: Corpus organisation in MAXQDA 11.

For example, we could simultaneously design two specialised corpora used for the study of different linguistic issues typical for legal language, i.e., a corpus encompassing legal documents, which can be further organised in sub-corpora, such as contracts, laws, regulations, etc., and a corpus encompassing specialised texts from the automotive industry, i.e., user manuals, technical specifications, manufacturing provisions, etc. Such corpora may be then individually investigated by each member of the project group in accordance with the specific research objectives, or they can even be contrastively analysed, if, for example, certain linguistic features are to be characterised contrastively in particular registers or even along certain time periods, i.e., time series.

By providing a user-friendly interface, MAXQDA reveals similar features to other Windows programs, facilitating quick and effective processing on behalf of the users.

After completing the design stage of a corpus, which, as previously indicated, offers innumerable compiling possibilities, we can embark on the linguistic analysis of the corpus by means of the drop-down menus and the various toolbars with buttons that offer quick access to the functions. It is worth mentioning that even at these stages, the investigation possibilities provided by the software are numerous. A general overview of a linguistic analysis model would imply the use of the following menus:

- the *Analysis Menu* – provides a series of analysis options applied especially to the lexical search and retrieval functions. Thus, the lexical search option enables the user to search within the document, or just in the activated document sets, memos, and retrieved segments. This function facilitates the search for certain words, phrases, or combinations. Also, the keywords in a context can be searched and automatically encoded. As indicated in the user manual, most of this menu functions relate to retrieval. We can chose various criteria for the segments to be found (e.g., OR, AND, logical combinations, or NEAR). Moreover, the retrieved segments can also be filtered based on certain criteria in the 'Retrieved Segments' window.
- the *Codes Menu* – enables the user to create and apply new codes on all the documents or only on the activated ones, or even to create a complete index of all codes assigned to all the document segments.
- the *Mixed Methods Menu* – is used to process and combine qualitative and quantitative data using documents and variables. Documents or document groups can be investigated based on the assigned variables, limiting the retrievals to certain document segments. The Quote Matrix and Crosstabs functions can be applied in order to indicate connections between the encoded segments and the selected variables.
- the *Visual Tools Menu* – enables the users to visualise the outcomes by means of seven different visualisation function options. *MAXMaps*, the tool for qualitative modelling; the *Code Matrix Browser*; the *Code Relations Browser*; and the *Document Comparison Chart*. The *Document Portrait* and the *Codeline* functions

also provide further visualisations which can be exported to Excel sheets as well under the shape of tables or graphics.

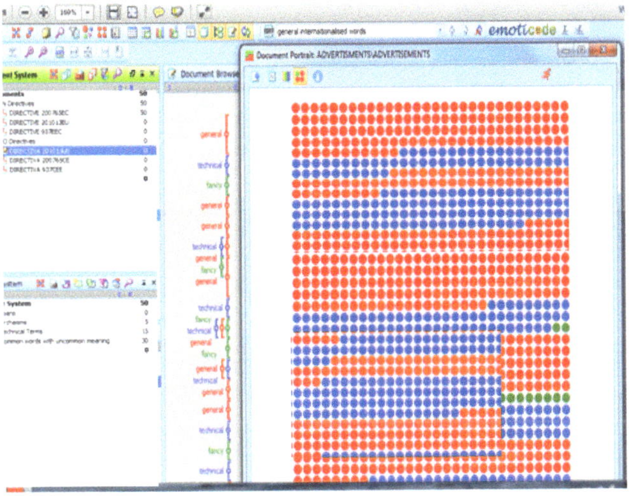

Figure 4.2: Print Screen: document portrait generation according to the assigned codes.

– The *MAXDictio Menu* – is an optional menu which offers a number of functions for quantitative content analysis, e.g., coding according to created dictionaries and viewing word frequencies.

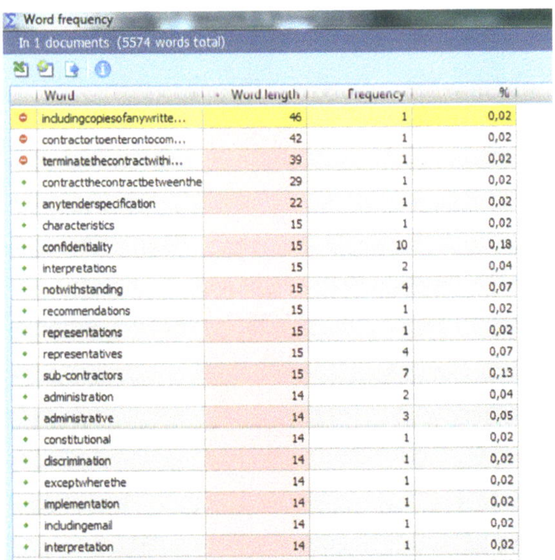

Word frequency
In 1 documents (5574 words total)

Word	Word length	Frequency	%
includingcopiesofanywritte...	46	1	0,02
contractortoenterontocom...	42	1	0,02
terminatethecontractwithi...	39	1	0,02
contractthecontractbetweenthe	29	1	0,02
anytenderspecification	22	1	0,02
characteristics	15	1	0,02
confidentiality	15	10	0,18
interpretations	15	2	0,04
notwithstanding	15	4	0,07
recommendations	15	1	0,02
representations	15	1	0,02
representatives	15	4	0,07
sub-contractors	15	7	0,13
administration	14	2	0,04
administrative	14	3	0,05
constitutional	14	1	0,02
discrimination	14	1	0,02
exceptwherethe	14	1	0,02
implementation	14	1	0,02
includingemail	14	1	0,02
interpretation	14	1	0,02

Figure 4.3: Print Screen: Word frequency list in MAXQDA 11.

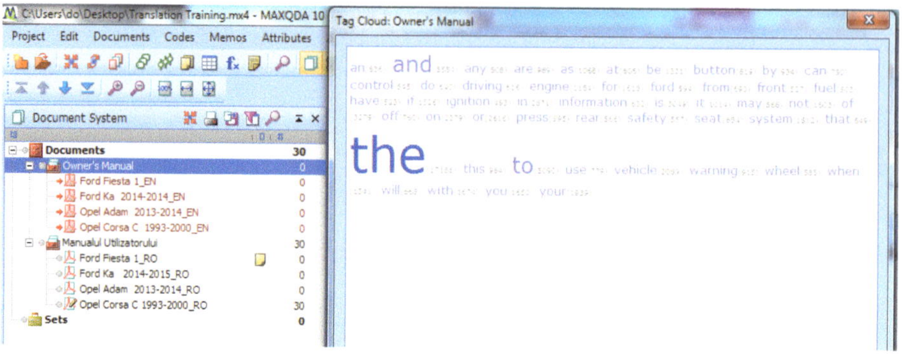

Figure 4.4: Print Screen: Tag cloud in MAXQDA 11.

As we can see, there are countless possibilities of analysis models, depending on the investigated linguistic issues and the specificity of the research fields applied. Of course, the design and implementation of simultaneous and multi-dimensional language analyses is possible, providing reliable and consistent outcomes in just a few minutes.

4.5 Conclusions

We can conclude that the computer-assisted approaches in corpus linguistics may lead, nowadays, to the refining and redefining of a wide range of theories of language. By means of dedicated software and computer-assisted tools, corpus linguistics has broadened its research directions considerably, smoothing the path towards new language explorations and theories.

As previously highlighted, within the context of advanced technologies, corpora are steadily exploited by tools that enable users to search through them rapidly and reliably. Most of such tools allow the production of frequency data, e.g., word frequency lists, document portraits, or comparison charts.

Unquestionably, there is a close link between the current status of corpus linguistics and modern technologies that brought to this field incredible speed, total liability, statistical reliability, sustainable results, and the opportunity to manipulate over considerably substantial and varied databases.

References

Aston. G. & Burnard, L. (1998). The BNC Handbook: Exploring the British National Corpus with Sara. Edinburgh: Edinburgh University Press.

Benskin, M. (1981). The Middle English Dialect Atlas. In Benskin, M. & Samuels L. (Eds.). So Meny People Longages and Tonges: philological essays in Scots and mediaeval English presented to Angus McIntosh. Edinburgh. xxvi-xli.

Biber, D., Conrad S., & Reppen, R.. (2006). Corpus Linguistics: investigating language structure and use. Cambridge: Cambridge University Press.

Francis. W. N. (1992). Language Corpora B.C. In Svartvik J. (Ed.). Directions in Corpus Linguistics: Proceedings of Nobel Symposium 82 Stockholm, 4-8 August 1991. Berlin: Walter de Gruyter, 17-35.

Kennedy, G. (1998). An Introduction to Corpus Linguistics. London: Longman.

Leech, G. (1992). Corpora and theories of linguistic performance. In Svartvik J. (Ed.). Directions in Corpus Linguistics: Proceedings of Nobel Symposium 82 Stockholm, 4-8 August 1991. Berlin: Walter de Gruyter, 105-123.

Lüdeling A. & Kytö, M. (2008). Corpus Linguistics an International Handbook. Vol.1. Berlin: Walter de Gruyter.

McEnery, T. & Hardie, A. (2012). Corpus Linguistics: Method, Theory and Practice. Cambridge: Cambridge University Press.

Scott, M. (2007). Oxford WordSmith Tools Version 4.0. Oxford: Oxford University Press.

Svartvik, J. (1992). Corpus linguistics comes of age. In Svartvik J. (Ed.). Directions in Corpus Linguistics: Proceedings of Nobel Symposium 82 Stockholm, 4-8 August 1991. Berlin: Walter de Gruyter, 7-17.

Online resources

Maxqda for Windows. Reference Manual. Berlin: VERBI Software. Consult. Sozialforschung. Available: http://www.maxqda.com [Accessed 2015, July, 30].

Part II: **Language and Translation: From Pen and Paper to the Electronic Environment**

Peter Sandrini

5 Towards a Digital Translation Policy

5.1 Introduction

Every human action takes place under constraints, with society as a whole, being the decisive determinant. To avoid confusion and disorder, a certain amount of structure and planning is needed: in this sense, a policy is focused on action, stating what is to be done and by whom; a policy states matters of principle, being an authoritative statement, made by a person or body with the power to do so.

Seen from this perspective, language also resembles a social phenomenon which may be addressed from various angles and disciplines. As such, it is a highly debated issue in society: 'Language issues are raised and considered in connection with cost, effectiveness, the language rights of minority groups, the democratic deficit and the need for a common European identity' (Fischer, 2007: 485). Many components need to be considered when structure and planning should be introduced into the use of languages in society: 'A host of non-linguistic factors (political, demographic, social, religious, cultural, psychological, bureaucratic, and so on) regularly account for any attempt of persons or groups to intervene in the language practices and the beliefs of other persons or group' (Spolsky, 2005: 2153).

Planning the use of language for a certain part of society – be it a minority, a territory, an organization, an institution or a company, requires well-considered rules and regulation, not just for determining the individual and collective use of languages in specific contexts, but also for setting up instructions on how to use the instrument of translation and all things that translation entails, among them translation technology.

The current paper focuses on translation as an integral part of a language policy and tries to sketch the contours of a digital translation policy in the sense of a translation technology policy as an essential component of a translation policy.

5.2 Language Policy

Language policy is a standard topic in linguistics which has seen a wealth of research interest in the 1950s to 1960s. It covers any 'deliberate attempt to change an individual's or community's use of a language or languages or a variety or varieties' (Kennedy, 2011) and was mainly subdivided into status planning and corpus planning. A language policy makes decisions about the status, use, domains, and territories of language(s) and the rights of speakers of the languages in question and, thus, governs mono- or multilingual communication in companies, organizations and institutions, even more so in a multilingual and multicultural society, where multilingual communication is common place.

There are four basic notions of a language policy as outlined by Spolsky (2006): 1) the division of a language policy into language practices, language beliefs, and ideology, and the explicit policies and plans resulting from language management or planning activities; 2) a language policy is concerned not just with named varieties of language, but with all the individual elements at all levels that make up a language (pronunciation, spelling, lexical choice, grammar, style, and bad language, racist language, obscene language, or correct language); 3) the domain of language policy may be any defined social, political or religious group or community (family, a sports team, neighborhood, village, workplace, organization, city, nation–state or regional alliance); and 4) a language policy functions in a complex ecological relationship among a wide range of linguistic and non-linguistic elements, variables, and factors.

Language has to be dealt with, intuitively or consciously: 'language policy exists even where it has not been made explicit or established by authority' (Spolsky, 2005: 2154). Such an implicit or informal language policy states a de facto standard without any written down rules or regulations, by adhering to covert practices. In most contexts, however, written down rules and regulations do exist and give explicit and formalized instructions on how language is to be used resulting in an explicit de jure language policy.

International organizations tend to limit the number of languages they use actively by choosing a limited number of official languages; supra-regional institutions also try to adhere to the principle of official languages; nation states often wish to guard their national language against the pressure from other language communities; linguistic minorities strive for language equality and preservation of their minority language against a majority language; multinational corporations set up strategic language policies to boost their international revenues, etc.

When a language policy is put in place, it may take the form of different types and outcomes: assimilation, non-intervention, establishment of a predominant official language, official bilingualism, strategic multilingualism, linguistic internationalization depending on the power relation between linguistic communities and the type of community or domain involved.

Multinational companies, for example, set up a strategic orientation with regard to language and culture in terms of ethnocentricity, polycentricity, or geocentricity (van den Born/Peltokorpi, 2010: 99) concepts which describe the degree of homogeneity, control, and local adaptation of business communication decisions. This not only affects marketing and sales, strategic orientation, but also human resource management and return on investment calculus in global companies (van den Born/Peltokorpi, 2010: 100).

A similar model reflecting internationalization and localization practices by global companies was presented by Lockwood (2000: 15) who distinguishes three approaches to global communication strategies: 1) the monarchistic approach with a strict top-down strategy, more or less following the ethnocentric model; 2) the anarchistic approach, a bottom-up model where all subsidiaries do what they want without any global strategy, similar to the polycentric approach, and 3) the federalist orientation, a mixed model with a central global strategy complemented by local adaptations, mirroring the geocentric approach.

The more linguistically oriented concept of language policy was abandoned by Spolsky in his following book (Spolsky, 2009) in favor of the more pragmatic notion of language management defined as an 'explicit and observable effort by someone or some group that has or claims authority over the participants in the domain to modify their practices or beliefs' (Spolsky, 2009: 4). A domain is 'distinguished by three characteristics' (Spolsky, 2009: 3): participants – social roles and relationships communicating, a location – physical reality of communication, and topic – the reason for speaking or writing. Spolsky lists some domain types, among them are the family domain, the religious domain, the workplace, the public linguistic space, the school domain, courts, hospitals and police, military language management, governments managing language and minority rights.

Translation may occur in all of these domains and it has to be managed or planned in some way. Unfortunately, translation has not been among the topics covered by a traditional language policy, nor is it touched upon by Spolsky's book (2009) on language management. Only recently, research on translation policy stirred the interest of researchers, in spite of it being a necessary component of language policy.

5.3 Translation Policy

In fact, Meylaerts (2009) makes translation policy an indispensable part of a language policy whether it is implicit, in the sense that it is not defined or written down, or explicit, i.e., written down or stated in a regulation: 'there is no language policy without a translation policy' (Meylaerts, 2009).

In the same online video, she defines a translation policy as the 'regulation of translation in official institutions' and applies it to linguistic minorities and their relation to the majority language. Three types may be differentiated according to Meylaerts (2009): 1) institutional monolingualism and non-translation, where translation is, at the same time, prohibited (for the direction from the majority language into the minority language) and obligatory (from the minority into the majority language) since only one language can be used in official documents; 2) institutional monolingualism and occasional translation, where translation is accepted as a granted exception, and 3) institutional multilingualism and multi-directional obligatory translation, which in most cases, is not feasible in strict application or different territorial levels of translation policy are applied.

In a later publication, Meylaerts (2011) extends the concepts of a translation policy to include international bodies as well for which analogous three types are identified: 1) non-translation or the use of a lingua franca in which case translation is still necessary on a personal level, i.e., institutional non-translation presupposes individual translation; 2) non-translation within the institution combined with translation for communication between the institution and the outside; and 3) multilateral translation when all languages are translated into all other languages.

Based on what has been said before about the general concept of a language policy, we may argue that a translation policy cannot be confined to official institutions and international bodies, it may very well be applied to multinational companies or international organizations as well. The very three types of a translation policy outlined by Meylaerts (2009) may be linked to the localization strategies of global companies, as stated by Lockwood (2000). Basically, all types of domains, according to Spolsky (2009), may be submitted to the management or regulation of translation in one way or another.

In this sense, a translation policy may be characterized by applying Spolsky's features of a language policy: 1) translation can be subdivided into translation practices, beliefs and ideology, and the explicit policies and plans resulting from translation management or planning activities that attempt to modify the practices and ideologies of a group of translators, meaning that, there always is some kind of theoretical address in translation, in the sense of an implicit or explicit statement about what is really done and how it is done; 2) translation policy is concerned not just with named varieties of translation and localization, but also with all the individual elements at all levels that make up translation; it can apply to the lexical choice, style, textual factors, decisional strategies in the sense of what is translated and what is not, and all levels of quality management, assessment, post-editing, etc.; 3) a translation policy operates within a community, of whatever size, be it professional – ethical codes of conduct by professional organizations, training institutions, translation departments – or voluntary translators – like Translators-without-Borders, The Rosetta Foundation, free software translators communities, fan translator communities and many more – or even a given social group like a regional minority or a refugee community, for example; 4) it is determined by a complex ecological relationship among many linguistic and non-linguistic factors and variables: that translation cannot be reduced to a purely linguistic activity that has been discussed in translation studies again and again for the last 40 years, where different influences on translation, such as components from sociology, action theory, information technology, communication studies, as well as cultural theory, have been analyzed.

To clarify the concept of a translation policy, we may apply the set of standard W-questions: who, what, when, where, why, in what way, by what means. This brings us closer to the substance and contents of such a policy.

- *Who* is going to translate? Should there be in-house staff or freelancers? Fundamental issues may be raised by this question, such as: Are multilingual individuals expected to translate, or is professional translation promoted?

Some linguistic minorities opted for the first, arguing that there is no need for professional translators when a proper language education is put in place. International organizations and companies chose professional translation by experience. Another basic question could be: Should domain experts translate or should translation be reserved for trained professional translators? This was and still is the object of intense debates between buyers of translation services and language service providers. Closely related to this, the following question arises: Where are translators coming from? Are there translators training institutions or plans to set them up?, etc.

- *What* is going to be translated? What kind of texts are going to be subject to translation according to the status of a text (official or non-official text), according to a specific domain (legal texts), to text types, media types, etc. What languages are translated into what other languages? What are the translation volumes produced by these decisions? And most importantly, what is not translated and what are the reasons for excluding translation in some cases?
- *When* is a text translated? Should translations be available immediately or is there a time gap between the publication of the source text and its translation corresponding to the necessary production time? In localization, the concept of 'simship' or simultaneous shipping of a product describes the synchronized release of a software product in all its language versions at the same time. In a few multilingual settings, different language versions of official texts are produced together by subject specialists and linguists at the same time. A time lag between the publication of the original version of a text and its translation could very well indicate an inferior social role or legal status of the target language; and vice-versa a synchronous multilingual publication could underline the equal status of the language involved.
- *Where* are translations produced? Do central translation offices exist where translation efforts are coordinated and translation resources are shared? Do specific institutions have translation offices that possibly pursue a different translation policy? Are there in-house translators, are freelance translators involved or is translation outsourced to translation agencies? What are the common guidelines and regulations for translation tenders?
- *Why* is something translated? Reasons for translation could be political ideology, minority rights, accessibility of official documentation, etc. This reflects the ideology behind translation as well as the legal and social status of languages.
- *In what way* is translation taking place? This involves translation strategies and methodology, translation management and efficiency, reuse of translations, quality assurance, revealing the overall organization of translation and its workflow.
- *By what means* is translation produced? This concerns, above all, the impact and use of translation technology, especially the use of automatic machine translation systems or semi-automatic translation environments, translation memory systems, multilingual corpus and terminology support for translation.

5.4 Digital Translation Policy

Today, in the age of globalization and after the 'technological turn in translation' (Cronin, 2010), we may say that there is no translation policy without a digital translation policy; i.e., no decisions about translation can be taken without taking into account the digital environment of communication and text production or the influence of digital technology on the production of translation. Translation Technology has become a core concept for modern translation practice and theory. Translators not only need to be proficient in translation technology, but they should master it, individually and as a community: 'In general, the way to advance within the profession usually involves more conceptual control over technology, not less' (Biau Gil/Pym, 2007: 19). Technology dominates the process of translation and 'only when translators are critically aware of the available tools can they hope to be in control of their work' (Biau Gil/Pym, 2007: 19). Translation can only survive as a profession and as an autonomous academic discipline if it integrates technology: 'La definición de un espacio profesional autónomo y digno supone no renunciar a mantener el mayor grado posible de control sobre los procesos de traducción' (Diaz – Fouces, 2011: 10).

The use of technology in translation presupposes planning and management to avoid uncoordinated deployment and usage. This is the task of a digital translation policy, a term which may be paraphrased as a translation technology policy stating the matters of principle on how to deal with translation in a digital environment and what is to be done and by whom. To achieve this, a coordination unit or decisional body on translation technology has to be appointed, if it is not already existing for translation in general, which has the political and administrative power to make authoritative statements.

There may be different degrees of such a decisional authority on translation technology which can be described by the three strategies outlined already for a language policy and a translation policy: the two extreme approaches, either strict regulation from a central coordination representing the top-down, monarchistic approach, or everybody does what she wants representing the bottom-up, anarchistic approach, or the third mixed or federalist approach with some central guidelines in conjunction with limited autonomy.

In analogy to the contents of a translation policy, the decisions a translation technology policy has to deal with may be described by asking the fundamental W-questions.

- *Who* is in charge of translation technology? Where lies the responsibility? With each individual translator, with every translation office, or with a central coordination unit? In addition, who takes care of installation, networking, and servicing of translation technology applications? Again, there is an option of a central coordination unit or a decentralized management.
- *What* digital text formats are dealt with as source and target texts? What kind of translation technology is adopted, MT, TM, etc.?

- *When* is translation technology integrated into the translation workflow? There could be some kind of pre-editing of source texts to adapt them to a particular machine translation system, or post-editing and quality assurance measures applied to target texts.
- *Where* is translation technology installed and made use of? Only in dedicated translation offices, in every administrative office that deals with bi- or multilingual texts?
- *Why* is translation technology used? Why is one kind of application preferred over another? What are the reasons for adopting it in the first place, meaning what are the particular benefits, and what the costs involved?
- *In what way* is translation technology adopted? Are there single desktop applications, one central web interface, etc.? Are commercial solutions or readymade free software applications put in place, or do they develop their own proprietary systems?

The motivation for adopting a translation technology policy corresponds more or less to the reasons for actually using translation technology: increase in productivity, consistency of texts and terminology, streamlining of translation process, exchange of linguistic data. In addition, with a sensible translation technology policy in place, a careful and balanced adoption, application and use of translation technology is possible, thus increasing the benefits mentioned above. Furthermore, a digital translation policy provides the basis for the allocation and distribution of human and financial resources to this purpose.

A well-thought-out digital translation policy also includes evaluation criteria (Gazzola, 2014) to check and monitor the allocation of these resources periodically by carefully weighing the advantages and disadvantages of alternative technology policy options. The two main factors of such an evaluation are efficiency and fairness in the sense of assessing who wins and who loses relatively (Gazzola, 2014: 2).

By having a look at the subdivision of translation technology (Quah, 2006: 42), we may envisage a translation technology policy with three specific sub-domains next to a more general aspect with the integration of translation into a digital workflow of text production and publication: computer-aided translation (CAT) tools policy, automatic or machine translation policy as well as a translation data policy, the first two being already devised by Quah (2006: 42).

5.4.1 Computer Aided Translation (CAT) Policy

A CAT policy encompasses all decisions about the adoption and use of computer-aided translation tools. This includes not only the decision which commercial or open source product is most suitable for the specific environment, but also how the chosen tool is adopted, installed, supported and evaluated. CAT includes translation memory applications, terminology management and term extraction tools, alignment tools, bilingual digital text corpora applications, quality assurance applications, localization tools, web localization tools – either autonomous or integrated into web content management systems, translation management tools, as well as the adoption of controlled language and standard file formats.

5.4.2 Machine Translation Policy

Machine translation may be adopted in two different ways: as an automatic translation service or in support of human translators. Both ways need careful planning; in the first case, machine translation systems can be customized to the specific requirements and made available to the general public, in the second case, it needs to be integrated into a translation environment tool like a translation memory system. Whichever way machine translation is used, you may choose between different types of systems: rule-based systems, statistical systems or hybrid applications.

Licensing may also play a role in adoption of MT systems: there are commercial systems or open source engines available. One example for the latter worth mentioning is the freely available Moses statistical MT engine developed with the financial support from the European Union; a platform that serves as a basis for many successful machine translation adoptions either out-of-the-box or as a customized system. When customizing a machine translation system, large amounts of bilingual text data are required and decisions are needed as to which text types, domains or languages should be chosen.

There is also a choice between offline and online system installations; there is even the possibility to use a general-purpose online system like Google Translate and integrate into CAT tools.

5.4.3 Translation Data Policy

Big Data has become a buzzword these days in almost all business domains and translation is no exception. Great amounts of linguistic data are produced by translators day by day in the form of translation memories, terminology files or bilingual texts. These data can be leveraged for future translations, thus increasing efficiency and productivity.

In this regard, we may state explicitly that there is no translation technology policy without a translation data policy. Planning and decisions affect the choice of data formats, storage options, licensing and accessibility of data, etc.

Reuse of linguistic data and dissemination of translations affect particularly Public Service Translation since this kind of translation aims at making texts accessible to minorities and smaller cultural groups; free access to these data facilitates new translations.

5.5 Conclusion

Speaking of translation technology, Diaz-Fouces (2011) postulates an integration of technology in all translation courses and says that 'esa transversalidad debería complementarse necesariamente con una apertura a la pluralidad de opciones, que lleva aparejada una imprescindible formación en criterios para elegir' (Diaz–Fouces, 2011: 14) (these transversal skills must necessarily be complemented by an opening up to the plurality of options which necessarily entails teaching selection criteria). This not only applies to translator training, but as well to translation practice in general. A policy is precisely about weighing different options by applying appropriate criteria, and the integration of such transversal decisional skills into translators training would not only complement the curricula, but introduce badly needed managerial or executive functions for translators as well. A translator with an academic background and a university diploma should have all necessary skills to be able to make autonomous and well-founded decisions regarding the use of translation technology.

Planning the use of translation technology is a necessity in all cases where translation fulfills a social function and more people are involved in translation. Translation technology has become rather complex and, thus, successful adoption demands careful consideration in each of the three subfields mentioned above. Once in place, translation technology must be monitored and evaluated periodically.

References

Biau Gil, & Pym, A. (2006). Technology and Translation (a pedagogical overview). In A. Pym, A. Perekrestenko, & B. Starink (eds.), Translation technology and its teaching: (with much mention of localization) (pp. 5-20). Tarragona: Intercultural Studies Group, Univ. Rovira i Virgili.

Cronin, M. (2010). The Translation Crowd. Revista Tradumatica, 08/2010, 1-7.

Diaz–Fouces, O. (2011). ¿Merece la pena introducir el software libre en la formación de traductores profesionales?. Presented at Language and Translation Teaching in Face-to-Face and Distance Learning, Universitat de Vic. Vic, 8 April 2011. Available at: http://www.academia.edu/3487697/_Merece_la_pena_introducir_el_software_libre_en_la_formacion_de_traductores_profesionales – accessed in April 2015.

Fischer, M. (2007). Language policy at European level. In J. Muráth, & A. Oláh-Hubai (eds.) Interdisziplinäre Aspekte des Übersetzens und Dolmetschens – Interdisciplinary Aspects of Translation and Interpreting, (pp. 485-498). Vienna: Praesens.

Gazzola, M. (2014). The Evaluation of Language Regimes. Theory and application to multilingual patent organisations. Amsterdam: John Benjamins.

Kennedy, C. (2011). Challenges for language policy, language and development. In H. Coleman (ed.), Dreams and realities: developing countries and the English Language. London: British

Council. Available at: http://www.teachingenglish.org.uk/article/dreams-realities-developing-countries-english-language – accessed in April 2015.
Lockwood, R. (2000). Have Brand, Will Travel. Language International 12/2. 14-16.
Meylaerts, R. (2009). No Language Policy without Translation Policy. On-line video: Available at: http://tv.uvigo.es/video/25506 – accessed in April 2015.
Meylaerts, R. (2011). Translation policy. In Y. Gambier, & L. van Doorslaer (eds.), Handbook of Translation Studies: Volume 2 (pp. 163–168). Amsterdam/ Philadelphia: John Benjamins.
Quah, C. K. (2006). Translation and Technology. New York: Palgrave Macmillan.
Spolsky, B. (2005). Language Policy. In J. Cohen, KT. McAlister, K. Rolstad, & J. MacSwan (eds.), ISB4 Proceedings of the 4th International Symposium on Bilingualism (pp. 2152-2164). Somerville: MA: Cascadilla Press.
Spolsky, B. (2006). Language policy. Cambridge: Cambridge UnivPress.
Spolsky, B. (2009). Language management. Cambridge: Cambridge UnivPress.
van den Born, F. & Peltokorpi, V. (2010). Language Policies and Communication in Multinational Companies Alignment With Strategic Orientation and Human Resource Management Practices. Journal of Business Communication 47, 97-118.

Mariana Pitar

6 The Impact of New Technologies on Specialised Translation

6.1 Introduction

Nowadays, almost any activity can obviously not be carried out without the support of computer science, either as an information source or an instrument. The boom in the domain of computer science and the omnipresence of the computer and of the World Wide Web in all activities does have an effect on the translation activity, not only in terms of the sources to use, but also of the translation tools. All these means are more and more numerous, so that their use replaces, for most people, the printed hardcopy sources. As long as the typing of the translation is done on the computer and as long as it is connected to the nternet, the direct search for online sources is easier and much more in the user's hand. A dictionary opened in a window in parallel to the text one works on represents the first research source for the words in the target language and also the easiest one to handle.

Even though they are the most frequently used, dictionaries are, however, not the only instruments from which a translator benefits from.

For a translator to carry out a good translation, he must be cognizant of the domain in general, of concepts as well as of the terms which cover these concepts in the source language and in the target language. The documentation research in libraries, the dialogue with a specialist, the research done in hardcopy dictionaries are not only time-consuming, but also insufficient activities for the translator. The Internet is more and more loaded with all sorts of information; all one has to do is find the right one. Yet simultaneously, it is increasingly necessary to select the information depending on one's needs and its quality.

In what follows, I am going to make a review of the instruments and sources provided by the World Wide Web in terms of information or as a free instrument for the translator's use. I have tried to choose the most interesting and useful data.

To classify it, I have suggested as a criterion the specific stages of specialised translation (Pitar, 2013) by highlighting the most interesting sources or tools. Without getting into details about the way software works or about a precise analysis of the existing offers, I aim to show the resources one can use in each of these stages.

6.2 Stages of Specialised Translation

6.2.1 Delimitation of the terms

A term must not be confused with a word. Terms may be made up of one or several words, but they can also be represented by all sorts of symbols. A brief classification of terms according to their form includes the following:

- simple terms – made up of one word only: *machine, nucleus, purify, etc.*
- complex terms (synapses) – made up of several words: *sewing machine, cover well, blue water gas, etc.*
- acronyms: *NGO, TGV, NATO, UN, etc.*
- codes, formulas: *A4, NaCl, etc.*

If we only take into consideration the terms made up of one or several words, there is a resulting problem concerning the limits of the term. A combination of words may be more or less fixed, more or less difficult for the translator to recognize. It depends on how specialised he is in the domain and on the difficulty of the domain envisaged. Some terms, especially in technical domains, are transparent enough; for instance, the names of machines or gadgets, which are usually formed of a hypernym term that is subdivided into a group of hyponyms. Here are some examples:

Connection
Terminal connection
Terminal connection diagram

Machine
Machine control
Machine vector control

Other terms are rather ambiguous and their delimitation differentiates between a term and its modifier, a term and a collocation or a phrase. In order for the translation to be accurate, the term must be well-delimitated, which implies a good knowledge of the domain.

6.2.2 Identification of the concept covered by the term

As I have already mentioned, good knowledge of the domain is a prerequisite for success. Before starting the actual translation, a translator must make a list of the domain-specific terms and ensure that he is aware of their meaning, i.e., the concepts covered by these terms. To define their meaning in a more precise manner, one must always think of the domain, even in the case of a simple term. Despite the claimed univocality of terms in specialised registers, one term may cover various notions depending on the domain. Thus, the translator must inform himself on the meaning of terms in the domain(s) of the text to be translated. Establishing relationships among these terms will lead to a good understanding of the domain.

6.2.3 Searching for equivalent terms in the target language

The actual translation begins by searching the specialised terms in the target language. The primary source for this activity is the specialised dictionary. We should remember that there is not always a dictionary available for all specialised domains and especially for the highly specialised subdomains. Often enough, such dictionaries include only the most general terms of the domain. Domain thesauruses and very specialised databases represent accurate enough information sources. The search in specialised text corpuses for both languages may be the final solution to find appropriate equivalents.

6.2.4 Translating the text

Once these stages have been completed and the nucleus terms are known, the next step is to fill in the textual canvas with the other words, collocations, expressions, and phrases typical of the specialised language of the domain.

6.3 Online Sources and Translation Tools

In what follows, I am going to mention the tools that I consider the most interesting for translators and that may help them in their endeavour.

One of the criteria employed in the choice of the products mentioned was, for certain tools, such as dictionaries, automatic translators or databases, the presence of the Romanian language as one of the two languages to work with, in particular, the target language.

Another criterion was that of the complexity of the analysed tool or software. That is why, I have not limited myself to the simple online dictionaries, which only give the first meaning of a word or term, without mentioning the sources, and whose quality is often questionable. Given the multiple functions certain tools include, their relation to a certain stage of the translation process has not always been straightforward. One such example concerns the form and functions of online dictionaries, which most often go beyond the simple search for an equivalent at word level and allow the translation of several sentences, even texts, thus also functioning as automatic translators. Sometimes, we deal with web pages offering several types of tools, which is why, I am going to analyse the categories of such tools, while also trying to keep track of the common theme announced above, i.e., the specific stages of translation.

One final and essential criterion was the free access to the above-mentioned sources, given that the target audience of this study is an independent translator and students as future translators.

6.3.1 Tools for extracting and delimiting terms

A criterion that helps in the delimitation of a term is its frequency. A research in the specialised texts of the domain may help us find not only the most frequent terms, but also collocations and contexts. It is the case of text analysers that have multiple functions. In this category, I have chosen term extractors and concordances.

6.3.1.1 Term extractors

Text analysers are automatic term acquisition tools, which employ a method of contrasting specialised and non-specialised corpuses with the aim of identifying terms. They allow the insertion of a text and return as a main result a list of candidate terms extracted from the text.

One such tool is *Termostat*, a term extractor available online, which includes linguistic and statistical methods and which considers the structure of units and the relative frequency of identified units in the analysis corpus. This software may also offer several types of term analysis:

- frequency statistics
- matrix
- spelling variants
- context
- structures
- the breaking down of the complex terms into headword and derivatives and compounds, as well as the relationships between the terms, and a chart suggestion.

Given these complex functions, such software may help in the conceptual organization of domains and represent a starting point in the making of databases.

6.3.1.2 Concordances

Concordances are simpler text analysers which allow the search for the contexts of a chosen word within a corpus. The contexts may vary in length depending on the user's choice. There are different types of corpuses, each such tool having its own texts. A concordance, in its most simple form, is an alphabetical list of the words in a text, together with the contexts in which they appear (Pincemin, 2010). Most often, words are provided as an index, where each word is centred in a standard-size window.

Our chosen example, *Le Migou* concordance comprises as a corpus the collection of the *Le Monde* newspaper, a medical corpus, a computer science corpus, and a Canadian-written press corpus.

Such software is very useful in the search for collocations, phrases, contexts and usage of terms.

In addition to the monolingual concordances, there are also multilingual concordances where the choice is made in texts written in several languages. In this manner, such a concordance may be used as translation support software.

WeBiText is a concordance available in several languages. We may choose the source language and the target language and use various corpuses, of which one exclusively uses the Canadian government websites. *WeBiText* is a multilingual translation support tool, which looks for translations of the words or phrases in the websites preselected and/or specified by the user, and used as bilingual corpus. The user selects the source language and the target language from the list of languages available and then chooses a site from the ones in the two selected languages. The tool allows for a research in more than 30 languages. The corpuses suggested depend on the selected pair of languages.

Another type of online text analyser – like *Textalyser* – makes statistical analyses by breaking down the text. Such text analysers provide information like the number of syllables, words, phrases, length of sentences, complexity and repetition of words. It is a useful instrument for translators or project managers in the field of translation.

6.3.2 Sources for equivalents

6.3.2.1 Databases and thesaurus

In order to understand the concepts of a domain, the translator must read equivalent specialised texts in the two languages – SL and TL. Equivalent information is offered by the databases that explain in detail the concepts and their connections and that also provide term equivalents in several languages. Thus, they function as mono- or multilingual dictionaries. Specialised explanatory dictionaries are most frequently found in the form of databases, which offer descriptive entries for each term, including a definition, the domain and/or subdomain, the relationships among concepts, etc., to enumerate just a few.

I am going to specifically refer to *IATE*, the database of the European Union. It comprises 8.7 million terms inserted by professional translators and it works in the 23 official languages of the EU in several domains: politics, international relations, law, economics, humanities, industry, agriculture, etc. The database has a user-friendly interface in a wide range of domains, with references, contexts and definitions. *IATE* contains all the information previously included in the former terminological databases of European institutions like *EURODICAUTOM* (Commission), *EUTERPE* (Parliament), or *TIS* (Council).

EuroVoc also belongs to the EU bodies. It is a multilingual thesaurus initially built especially for the processing of documentation of European Union institutions. It covers wide enough domains to include not only Community aspects, but also the national perspectives, with a focus on the parliamentary activity. The objective of the thesaurus is to provide information management and information distribution services with a coherent indexing tool so that they can efficiently manage their documentation content and allow users to conduct documentary research by using a controlled language. As any thesaurus, it has the advantage of graphic and symbolic representation of the relationships among concepts, which it can show graphically.

This clear presentation of the relations among concepts enables a better understanding of the domain and the avoidance of overlapping with other terms of the same form, but covering different concepts in different domains.

Another database is *Le grand dictionnaire terminologique* (GDT), a data bank of terminological entries of the Canadian government, with millions of terms in several languages in various domains.

Databases, thesauruses, glossaries may be grouped according to the domain, the language, but they are also grouped around several bodies like EU, UN, ECHA (European Chemicals Agency), etc.

Some very large databases and thesauruses represent genuine terminological portals because they include terms from several databases.

Agrovoc is a multilingual agricultural thesaurus belonging to the UN. The concepts can be visualized as tree structures with very complex hierarchical relations. In the case of plants and animals, these relations show the entire complexity of the biological taxonomy. The terminological entry appears in the same window as a part of the tree structure to which the term belongs. A click on one of the terms in the tree structure allows us to navigate to other superior or inferior related concepts.

Termscience is a multidisciplinary terminological portal. It is an enormous database and thesaurus, which includes the terms of other databases from various scientific domains. There are hundreds of thousands of terms, and the tree structures have numerous nodes. An advantage is represented by the complex search modes either in an alphabetical list or in the domain's tree structure. The alphabetical list is shown as a *cloud*, where the most frequent terms (sometimes appearing tens of thousands of times and which represent the root nodes of at least one tree structure) are highlighted by the use of a different colour and font size. This list may be shown in two manners: simply alphabetically or as a tree structure, depending on the term frequency. In addition to this, there is a direct search in the databases of origin, which are to be found in a scroll-down list.

6.3.2.2 Dictionaries

The specialised source for the translation activity has always been the dictionary, which, like databases, offers not only information on the terms, but also equivalents in several languages. Electronic dictionaries, as compared to the printed dictionaries, have several practical advantages, such as the search speed, the fact that the translator does not need to move in order to look for a dictionary because he can look words up directly from his workstation, the access to data, which in hardcopy dictionaries is more difficult to find. The electronic dictionary offers the possibility to perform a 'full text' search, which returns much more examples. According to Druţă (2009 : 535), 'Terminological dictionaries, including the electronic ones, are efficient instruments in promoting national terminologies and connecting to the international usual terminology. An electronic dictionary of terms, compiled according to modern exigencies, is an excellent source in learning specialised languages and in efficiently transferring knowledge from one culture to another and will facilitate interlinguistic and intercultural dialogue'.

The Internet offers a wide range of dictionaries. Even the term *dictionary* begins to undergo changes in meaning, as an increasingly larger number of dictionaries do not only provide the equivalent, but they have become actual translation software. The user may choose between the search for a term and the translation of a phrase, a sentence, or a text. The results are, of course, debatable, but it is not my aim here to carry out such an analysis.

Instead, I am going to show some dictionary types, which offer various and useful information that I have found interesting.

First of all, it is important to mention the websites that offer dictionaries for all languages. The most important ones here are *Lexilogos* and *Lexicool.* I suggest a synthetic classification of these dictionaries into the following categories:

- monolingual explanatory dictionaries
- simple bilingual dictionaries or with an integrated automatic translator
- simple or visual specialised dictionaries

Among these dictionaries, increasingly numerous, I am going to mention some that I find representative.

Sansagent seems to be the most important of the explanatory dictionaries, as it refers to several types of dictionaries – explanatory, of synonyms, analogic, etc.; it offers explanations for phrases and expressions, as well as the meaning in the specialised domains. It is also a bilingual dictionary and has Romanian among the languages available.

Glosbe and *Linguee* are both multilingual dictionaries with a large number of combinations of languages and domains. They are also translation tools, which combine a typing dictionary and a research engine, comprising hundreds of millions of bilingual texts. They offer a wide variety of translation examples with contexts stocked in the translation memories whose sources are mentioned. The principal corpus is formed of EU texts. In addition, audio recordings provide the pronunciation of the word. Of the two, Glosbe has the advantage of the number of languages available, which includes Romanian.

Visual dictionaries represent a special, more recent category. Their advantage lies in the illustration of the concept by means of pictures. They are especially useful for terms designating objects and their components. The domains are diverse enough and the definition given under each term helps its understanding. For the moment, they are rather limited in number. The most important ones are *Le dictionnaire visuel* and *Ikonet*.

Certain thesauruses belong to the same category of visual terminographic products. The most important one is the online *Visual thesaurus*, which uses all the multimedia elements, in particular sound and animation. The terms are shown on an interactive map, which changes as you select one of the terms in the tree structure. In the same window, by clicking on the terms, we find out their definitions and various types of relations they have with other terms in the tree structure: synonyms, antonyms, generic and partitive relations, associative relations, etc. This product is a model of interactive completion of other online products.

6.3.3 The translation proper

After all these preparatory stages for the translation, the translator may move on to the actual translation. He may use translation support software, such as automatic translators or translation memories that allow for an automation of the process and increased profitability.

6.3.3.1 Translation software

This refers to automatic translators. Created as a result of studies blending computer science, cognitive psychology and linguistics, these types of software attempt to give translation variants based on the analysis of the source text in terms of lexis, semantics and syntax, as well as the choice of terms and of syntax appropriate for the target text. The recent successes obtained in this domain over the past few years are remarkable indeed, but the results can always be improved. There is a wide enough range of products. These tools allow for an automation of the translator's work, but they need human intervention throughout the translation process. In this category fall two types of products: free ones, with average performances, and subscription-based ones, with better performances. The same product may exist in a free (or demo) version and in a subscription-based one.

Among these products, I remind of *Reverso*, *Systran* (fairly good in its professional version), *WorldLingo*, *Babylon*, *Google*, etc. I have also noticed ImTranslator, which includes Romanian among the language combinations and uses three translation services: *Babylon Translator, Google Translator* and *Microsoft Translator*. It also suggests in parallel three translation variants.

Automatic translation services, like *Google Translate,* are becoming better and better as a result of the insertion into their memories of improved translations sent by users.

I must also add here the more and more frequent function of the automatic translator added to online dictionaries.

6.3.3.2 Translation memories

If, in the case of an automatic translation, the computer is responsible for the translation of the text we insert, in the case of a translation memory, the computer gives the canvas, i.e., the software which is going to carry out two important operations. Firstly, it will align the source text and break it down into fragments. Then, after a textual analysis, it will search in the translations already existing in its memory and perform an automatic translation of the words, phrases, text fragments and even entire paragraphs, offering them as solutions for the translation. In this way, the precision of a new translation depends on the accuracy of previous translations stocked in the memory. The translator is, thus, responsible for the quality of the computer-provided translation.

A translation memory has the advantage of preserving all the previously completed translations, with a view to using them again and to sparing the translator from translating the same sentence several times (Peraldi, 2010).

The more the translations stored in the translation memory are similar in form and content, the higher the probability to find equivalent terms that have already been translated. As most often, a translator amasses texts from various domains, for better productivity in the translation activity, we recommend the building up of several translation memories, according to domains and even sub-domains. The texts made up of fragments that have a higher degree of repetitiveness, such as extended phrases of certain form-like documents, are very likely to find equivalents in the translation memory, up to 80–90%.

Among the best known translation memories, I would like to mention *WordFast Anywhere*, an online version, *Déjàvue* and *Omega T.*

In its online version, *WordFast Anywhere* has the advantage of including several research engines, which from the very beginning offer translation variants. It, thus, combines the advantages of an automatic translator with the advantages of translation memories.

I would also mention *Trados* among such products, which is a complex platform offering a translation memory and a management programme for databases. It is a professional tool and is available only in a subscription-based version.

6.4 Conclusion

As we have seen, the translator has the opportunity to improve his work by using sources and translation support software provided on the Internet or by professional agencies. Given all the software, the translator can nowadays use in his work, we may speak rather of a computer-aided translation (CAT), as the translator adapts all these tools to his working method. He must choose his own sources and tools according to the domain, the software performances, and the languages used. These facilities offered by computer science cannot and should not be ignored (Frérot, 2010). On the one hand, their performances improve almost overnight and they become increasingly easier to use; on the other hand, we must bear in mind that many translation offices require the translator to work with certain software. A translator must know how to work at least with one translation memory and draw up his own database, his own automatic glossary.

We observe the increasingly pronounced numerical outline of many of them, which are no longer simple electronic versions of print products, but which include more and more multimedia elements and become increasingly interactive.

All these instruments are time-saving, contributing to the profitability and increase in quality of the translation. They do have advantages and drawbacks and regarding the *when* and *how* of their use, the last word is, of course, the translator's.

As concerns the importance of presenting such information tools to students/ future translators, we believe it essential that they become familiar with such state-of-the-art technological advancements in the domain. Given the growing number of such tools online, more and more developed and complex in terms of information, students must be encouraged to search for and discover the new tools, try them, establish a value list, and include them in their translation activity.

References

Druţă, I. (2009). Modalitati de transfer al cunoştintelor în societatea informaţionala prin dicţionare terminologice electronice. In The proccedings of the EUROPEAN INTEGRATION-BETWEEN TRADITION AND MODERNITY Congress, vol 3., Târgu Mureş : Editura Universităţii „Petru Maior".

Frérot, C. (2010). Outils d'aide à la traduction: pour une intégration des corpus et des outils d'analyse de corpus dans l'enseignement de la traduction et la formation des traducteurs. In Les Cahiers du GEPE 2/2010 _ [accessed 20.07.2015]

Pitar, M. (2013). Manual de terminologie şi terminografie. Timisoara : Mirton.

Péraldi, C. (2010).Traduction assistée par ordinateur: entre théorie et pratique. In Les Cahiers du GEPE 2/2010 _ [accessed 20.07.2015]

Pincemin, B. Concordances et concordanciers. De l'art du bon KWAC. In Texto _ [accessed 20.07.2015]

Sources and software cited:

http://blog.atenao.com/traduction-professionnelle/liste-des-logiciels-de-traduction-assistee-par-ordinateur-174 [accessed 20.07.2015]

https://www.youtube.com/watch?v=fvFjLQLTmM8 [accessed 20.09.2015]
https://glosbe.com [accessed 20.09.2015]
http://www.infovisual.info [accessed 20.09.2015]
http://www.ikonet.com/fr/ledictionnairevisuel/ [accessed 20.09.2015]
http://www.wordfast.com/ [accessed 20.09.2015]
http://iate.europa.eu/ [accessed 20.09.2015]
http://www.visualthesaurus.com [accessed 20.09.2015]
http://www.termsciences.fr [accessed 20.09.2015]
http://aims.fao.org/vest-registry/vocabularies/agrovoc-multilingual-agricultural-thesaurus [accessed 20.09.2015]
https://translate.google.com/ [accessed 20.09.2015]
http://freetranslation.imtranslator.net/ [accessed 20.07.2015]
http://www.lexilogos.com/ [accessed 20.09.2015]
http://www.lexicool.com/ [accessed 20.09.2015]
http://www.sensagent.com/ [accessed 20.09.2015]
http://eurovoc.europa.eu/ [accessed 20.09.2015]
http://www.granddictionnaire.com/ [accessed 20.09.2015]
http://textalyser.net/?lang=fr [accessed 20.08.2015]
http://termostat.ling.umontreal.ca/ [accessed 20.07.2015]
http://olst.ling.umontreal.ca/migou/ [accessed 20.09.2015]

Felix Nicolau

7 The Transfer of Signs between Heterogeneous Systems: Incongruent Equivalences

7.1 Introduction

'The relationship between a word and what it stands for is like the relationship between a flea and a dog. If the dog is healthy, so is the flea. If the dog dies, so does the flea – unless it's able to find another dog' (Metcalf, 2002: 162)

It may have been the word in the beginning, but what the divine word represented we do not know for sure. God created the universe with the help of a language. Were there only words or also other kinds of signs? What kind of language could have been involved in the process of pronouncing and creating at the same time? Our language has lost this impact and we can notice that technology weakened even more the seductive capacity of our language. In order to arrest people's attention, we resort more and more to a mixture of text and image.

With the advent of Translation Studies as a distinct discipline, it seems that additional emphasis is put on the conditions and implications of the process of translation. The problems of transfer and equivalence make progressively room for the in-betweenness. As the interlingual transfer cannot be set free of the implications of power and hegemony, some theorists consider that 'all translation may be said to be indirect speech, in as much as it does not repeat the ST, but reformulates it' (Hervey, 2000). They refer especially to the *gist* translation taking place in intralingual contexts. The gist translation would interpret the message and can be seen as the compressed form of the *exegetic* translation. In both cases, we have a process of rephrasing, which involves both translation for the gist and exegetic translation.

7.2 The Ever-surprising Intersemiotic Translation

As we know, the saga of intersemiotic translation, involving an almost infinite semiosis, began with Roman Jakobson's seminal essay from 1959, *On linguistic aspects of translation*. Intersemiotic translation is a way of bypassing the semiotic system of language. The Russian–American linguist defines this less common type of translation as a translation of a verbal sign into a non-verbal sign. In the same paper, he postulated that only interlingual translation is 'translation proper' (Hatim & Munday, 2004). In the meantime, Translation Studies has evolved dramatically, including nowadays audio-visual translations with their sub-strata: sign language, intralingual subtitles, lip synchronization for dubbing and interlingual subtitles. There is no barrier left between linguistic signs and non-linguistic signs. Thus, the semiotic system of pure language becomes almost outdated. In today's global culture, a system of semiosis with mixed registers of communication is becoming the dominant approach.

Translations include nowadays *pictorial* and *iconic-linguistic* registers. In 1994, W. J. T. Mitchell (*Picture Theory: Essays on Visual and Verbal Representation*) analysed terms like *iconicity* and *image text*. The accent fell on the participatory intermediality, able to open infinite possibilities for the condition of translatability, owing to the fluency of mirrored and intertwined iconicity. But, the prejudiced views took advantage of the consecrated definitions of a sign. So, what entities would qualify for the status of a sign? Petrilli (2003) considered a few distinctions:

> *[T]rans-, inter-, dia- are prepositions and prefixes that specify the modality of being of the sign, the sign process, semiosis [...] Semiosis is a transsign process, an intersign process. Something that is not capable of relating to something else that signifies it, utters it, translates it, interprets it, responds to it, is not a sign.*

In order words, are pictures, paintings, gestures, and sounds signs? Can they be the subject matter of translation theorizing? Taking a further step onward, Torop (2004) asserted the idea of a 'partial overlap of signs and languages or sign systems of different arts'. Transfer, transmission and exchange are understood now on a larger scale than ever, and Roman Jakobson, in spite of the subordinating intersemiotic translation to the 'proper' translation between languages (interlanguage translation, implying texts) has an undisputable merit for having recognized the possibility of translating non-verbal messages: 'Intersemiotic translation or transmutation is an interpretation of verbal signs by means of signs of nonverbal sign systems" (Venuti, 2004). He also provided examples of transmutation paradigms from verbal art to dance, cinema, music or painting. This overlapping equated to transmutation is the mark of incongruity and heterogeneity. So far, the transfer of signs has been admitted only between homogeneous systems. Once the linguistic stage of Translation Studies got dated, intersemiotic translation raised the challenge of translating not only verbal signs into non-verbal signs, but also categories of non-verbal signs into other categories of non-verbal signs. Already Greimas (1966) perceived the emerging dynamics of intersemiotic translation: 'every signifying totality [ensemble] which is by nature different [has a different character] than natural language can be translated, more or less accurately, in any given natural language". The problem of accuracy looms ominously in the background, but we have to embrace a permissive attitude towards a nascent modality of universal communication. Umberto Eco (2001) also assented to such a tolerant approach when he considered translation a metaphor, a transference or an adaptation.

7.3 Media Literacy and the Visual Fluency of Messages

If texts proved to be in strict relation with their cultures of provenance, media theory admitted to multimedia signs being determined by the screen they appear on. Actually, as our society re-turns towards an image-informed way of communication, it is more and more difficult to practise an abrupt differentiation between a text and image. Communication seems to maintain its fluency even when it is realized with non-linguistic means. Media literacy tends to become more attractive than linguistic literacy. In these conditions, intersemiotic translation acquires strategic importance. Signs and images conjure each other into an amalgamated perceptual medium. Monosemiotic texts are a reality of the past in terms of impact force. In order to get behind the struggle between words and pictorial signs, and in this way to enhance the strength of the communicative function of translation, we should be open to the synergy resulted from the combination of different semiotic systems. This type of intersemiosis was called *multimodality* by Kress and Van Leeuwen (2001). The new perspective upon Translation Studies engages the acceptance of a multi-levelled reality with transdisciplinary openings.

In advertising, for instance, the verbal message can be used as an iconic system. This is what Guillaume Apollinaire did in *Calligrammes* (1913), wherein, among other *pattern* poems, he included a poem in the shape of the Eiffel Tower. The layout of words, in such a case, is more important than their innermost meaning and advertisers of a later date speculated on this experiment.

(https://wordsandeggs.files.wordpress.com/2008/10/250px-guillaume_apollinaire_calligramme-1.jpg)

Figure 7.1: Guillaume Apollinaire's poem on Eiffel Tower.

In other cases, the Eiffel Tower is represented like an adorned Christmas tree, which, using the rhetoric switch of metonymy, symbolizes the charm of Paris during winter holidays. This approach proves that the devices applicable to texts achieve the same function in the realm of images. The overarching principle is that of intertextuality: 'the notion of intertextuality refers to close relationships of content and/or form between texts. No text stands on its own. It is always linked to other texts' (Martin & Ringham, 2006). By assuming the principle of *intertextuality*, intersemiotic translation resorts to transmutation engineerings which are highly creative and even critical. In the case of poetry translation, the intertextual awareness is compulsory, as the translator will decode and re-encode cultural-bound concepts, besides prosodic subtleties.

If the ideal transposition and rephrasing were impossible even in the first half of the 20th century, the less likely will they be in the multimedia communication epoch. Inside environments that make use of sensory words, vision, audition, and touch, all texts become 'polysemiotic multi-signs' (Gambier and Gottlieb, 2001). Multimedia translation is about drafts and screen translation, which include transitory, non-finite content. Additionally, the interconnection of the media relies on flexibility and teamwork. In drama translation, for instance, the text is preserved in a fluid and negotiable state, which can be modified at every rehearsal. Besides, in this field, translators have to cope with incessant revision. The stages of a negotiable translation, wherein translators are supposed to collaborate with other specialists, can be traced with the help of contrastive analysis and contrastive stylistics. This is the procedure of *parallel texts* or *comparable texts* (William & Chesterman, 2002). As Translation Studies has arrived to be 'informed by a babel of theories' (Kuhiwczak & Littan, 2007), we no longer should remain limited to the predominance of the linguistic code. The importance of signs transfers between verbal and non-verbal codes, through the mediation of hybrids like *iconotexts*, is increased by the upheaval of technological innovations. In these conditions, we are bound to accept that communication through translation is realized with the help of complex cultural scaffoldings. The illusion of transparency in a perfectly achieved translation, as Lawrence Venuti appraised it at the end of the 20th century, depended on fluency and it discarded 'translates', 'translationese', 'translatorese', 'jargonisation', slang, pidgin, Britishisms in American translations, and Americanisms in British translations (when it comes to English, of course). A fluid translation would avoid a 'doughy' or idiomatic syntax, whereas it would conserve the rhythmic sense and an expressive closure (not a 'dull thud') (Venuti, 2004). The point was to attain a naturalness of the transferred message as if it had never been displaced from its original context. Obviously, this approach referred to literary translations, as in technical texts, what matters is not the stylistic factor, but the accuracy of the context renderings. The freight trains analogy resists the passing of time: it is not the order of the cars that matters, but the intactness of the cargo (Landers, 2001).

7.4 Culturemes and Program Music

There are always situations and contexts in translation and nobody could be so lofty as to provide an all-encompassing-all-solving approach. We may oscillate between domestication and foreignization, between dynamic and semantic equivalences, between the skopos theory and the theory of translational action and finally to become supporters of the functional anthropophagic (or 'cannibalistic') approach in Brazil or of the *Neuorientierung* in Germany.

If the meaning of a sentence is always dependent on and connected to the syntactic net that surrounds it, then the meaning is an occasional creation. Translation too is circumstantial, which means that objectivity varies in function of certain parameters. Objectivity borrows subjective traits, whereas subjectivity tends to objectivity in order to legitimize itself. This situation becomes poignant when the translator has to deal with *culturemes*, defined as a phenomon existent in 'culture X but not present (in the same way) in culture Y' (Gambier, 2001).

Intersemiotic translations are less culture-proof than interlingual translations, as they resort to a larger range of signs. Let's think of *program music*, which is instrumental music carrying an extra musical meaning. With the help of sounds, composers translate ideas, legends, and literary plots of the subjects of paintings. This is exactly why purist musicians underrate this genre of music as being impure. In their opinion, only abstract or absolute music would transmit the genuine musical feeling. The program music at its best soared in the Romantic era, from Beethoven to Richard Strauss. But absolute music is utopia. In fact, every score reflects some extra musical 'content': series of images, moods and states of mind. The symbolic and evocative substance of music can be detected in a *siciliana*, for instance, which is a composition built on the rhythm of an Italian dance and which triggers associations of tranquillity. Beethoven's *Symphony No. 6* (*Pastoral*) contains examples of literal 'tone painting' (a bird call in the second movement). Descriptive elements are found in Japanese *samisen* music, with stylized sounds of falling rain and snow. Georg Friedrich *Händel* evoked the ravaging effects of plague in his oratorio *Israel in Egypt* (1739). It was Beethoven who unified the movements of a symphony or sonata into a psychological whole. In his *Symphony No. 3 (Eroica),* he contrasted states of mind and explored the transition between them. The unification of tendencies was practiced by Robert Schumann, who conjoined opposite phrases in *Carnaval*; the same method was applied in overtures by Beethoven (*Leonore* No. 3) and Felix Mendelssohn (*The Hebrides*). Carl Maria von Weber's *Konzertstück* (1821) and Hector Berlioz's *Symphonie fantastique* (1830) were both accompanied by a printed synopsis of the 'plots' at concerts, this gesture disclosing an overt programmatic attitude. This trend continued all along the 20th century with experimental composers like Alban Berg, Gustav Hols, Mike Oldfield or Vangelis. Apart from making use of leitmotifs, these composers – be they classical or popular – involved various programmatic elements in their compositions, like art rock, ambient music, space music, surf rock, jazz fusion, progressive rock, and new age. Thus, music acquires the ability to transmit conceptual and narrative messages in combination with sound effects. If 'pure' music engenders sensations and feelings, program music stimulates recognition and interpretation as mental operations specific to the translation process too.

7.5 The Intersemiosis of Concrete Poetry

Another situation of intersemiotic transfer and interpretation is that of concrete poetry, sometimes called 'shape poetry'. Its visual appearance is supposed to match its topic. Consequently, the words are disposed in shapes that illustrate the poem's subject both as a picture and through its literal meaning.

'The Mouse's Tale' from Lewis Carroll's Alice's Adventures in Wonderland is a pun on the words *tale/tail* and its words flow along a slithering line that gets thinner and thinner down to an ending point, imitating the shape of a tail or a snake. Indeed, the poem is rather snaky as it relates in a humorous tonality about dictatorial and murderous instincts. The translator, in this case, has to stay double-focussed: on the one hand, the layout requires flexibility in lexical selection, and on the other hand, the meaning must preserve its contradiction to the jocose shape of the poem. To the complexity of the operation, we have to add the exact reproduction of some rhymes like 'jury' – 'fury'.

The Mouse's Tale

Fury said to a mouse,
That he met in the
house, 'Let us
both go to law:
I will prosecute
you.-- Come, I'll
take no denial;
We must have
a trial: For
really this
morning I've
nothing to do.'
Said the mouse
to the cur,
'Such a trial,
dear Sir, With
no jury or
judge, would
be wasting
our breath.'
'I'll be
judge, I'll
be jury,'
Said cunning
old Fury:
'I'll try
the whole
cause, and
condemn
you
to
death.'

(http://sparkleteaching.blogspot.ro/2011/05/free-form-poem-mouses-tale.html)

Figure 7.2: 'The Mouse's Tale' from Lewis Carroll's Alice's Adventures in Wonderland.

The Brazilian poets of *Noigandres* group issued a manifesto in 1950, stating that concrete poetry is characterized by the congruence between its structure and content. This type of intimate reflection is achievable in various ways, one of them being the filling up of an outline shape related to the topic of the poem, as is the case with *snowman*:

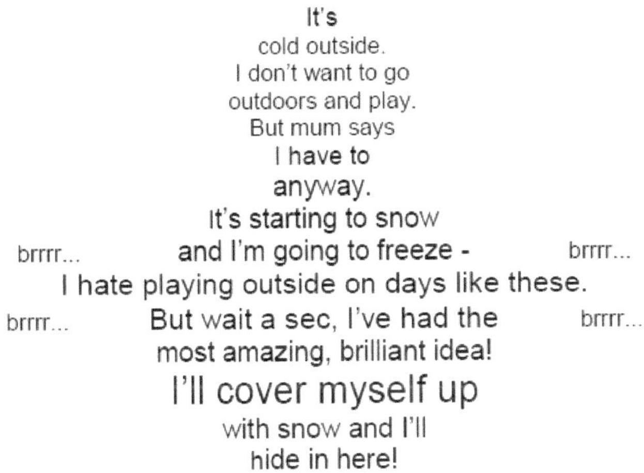

(http://www.poetry4kids.com/blog/news/how-to-write-a-concrete-poem/)
Figure 7.3: *Snowman* poem.

The outline can be gradually re-arranged during composition in order to obtain the desired image. This strategy anticipated the effects that can be created in some computer programs, like *Flash*. Consequently, the 'manufacturing' of concrete poetry consists of using the lines of a poem to make the blueprint of a drawing. The representation of the phenomenon of a growing picture looks like this:

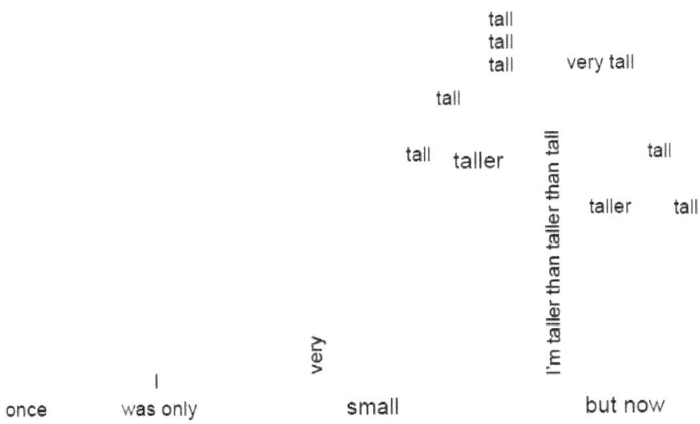

(http://www.poetry4kids.com/blog/news/how-to-write-a-concrete-poem/)
Figure 7.4: Concrete poetry.

Again, during the draft stage, we can make our writing bigger or smaller to get it all fit. In the end, we simply erase the drawing lines and are left with just the words in the poem to create the desired image.

Among the ancestors of concrete poetry, we find *altar poetry*. This type dates back to the ancient cultures of Persia and Greece and was lost until the 16th century when it reappeared in Europe. Altar poetry has the ambition to reflect the textual meaning through its arrangement on the page. The metaphysical poet George Herbert (1593–1633) replicated a wing in Easter Wings and created the sensation of flight 'on the back' of a lark, which, in its turn, symbolized the fervent, religious soul. The poem reinforces its meaning of rise and fall with the help of a suggestive, shape which is not only a mimicking reflection of the content, but also an extension of undertones:

Lord, who createdest man in wealth and store,
Though foolishly he lost the same,
Decaying more and more,
Till he became
Most poore:
With thee
O let me rise
As larks, harmoniously,
And sing this day thy victories,
Then shall the fall further the flight in me.

(http://www.poetryfoundation.org/poem/173626)

Figure 7.5: George Herbert's Easter Wings poem.

Later on, poets like E. E. Cummings (1894-1962) or Dylan Thomas (1914-1953) practised the *geometric* poetry. Although they enlarged the range of their inspiration, they observed the principles of altar poetry. Here, it is a poem representing an hourglass from *Vision and Prayer* by Dylan Thomas. In accordance with its layout, the poem showcases the fragility of birth and the exhaustion of death. The geometry of such a poem adds to the inner meaning and when a translator renders it into another language, s/he has to recreate the allusions related to content and shape, actually a double-range set of allusions.

Who
Are you
Who is born
In the next room
So loud to my own
That I can hear the womb
Opening and the dark run
Over the ghost and the dropped son
Behind the wall thin as a wren's bone?
In the birth bloody room unknown
To the burn and turn of time
And the heart print of man
Bows no baptism
But dark alone
Blessing on
The wild

(http://www.webexhibits.org/poetry/explore_21_visual_examples.html)

Figure 7.6: Dylan Thomas's poem – Vision and Prayer.

Dracula: Ballet and Intersemiotic Translation

The mirroring and recreation of one language into another one, part of a different system, is noticeable in the ballet Dracula. In 1999, David Nixon created the choreographed version of Bram Stoker's novel for Ballet Met and recreated it for Northern Ballet in 2005. Nixon kept the women characters alive *en pointe*, whereas those of the other world danced in flat shoes to suggest they need earth to survive. The director based his ballet on the novel, but the narrative substance is disposed in contrasts. There is a continuous play with light and shadows, indicative of the fact that living and dead need each other. The *pas de deux* between Dracula and Mina discloses the sensitivity of the vampire and maybe it is on account of this emotional quality that Mina forces him to bite her and make her the same as him: immortal, but monstrous.

The sets and costumes place the ballet in the romantic gothic style of the Victorian epoch. This historical distance would allow the public to focus on the exploration of human nature rather than on the bizarreness of details. The designer of the set was Ali Allen, who had already collaborated with Nixon on several intersemiotic productions, including *Wuthering Heights*. Thus, the set design provides a gothic backdrop for the ballet, which is often minimal, resorting to pieces of furniture to locate the scene.

(http://dancetabs.com/2014/09/northern-ballet-dracula-leeds/)
Figure 7.7: Dracula as a lizard.

In Act I, Dracula appears climbing down the side of a building as if a lizard, which reaffirms his dual nature: half man, half beast. The serial changes in the colour of the set reveal emotions and states of mind. In Act I, when the backdrop of pillars creates a screen that is coloured from neutral to blood red, we understand that Lucy has fallen under Dracula's spell.

(http://dancetabs.com/2014/09/northern-ballet-dracula-leeds/)
Figure 7.8: Dracula puts a spell on Lucy.

Tim Mitchells was responsible for the lighting design. He also had worked on David Nixon's Hamlet ballet version. The contrast of light and shadows was created through the use of blocks of light, side lighting, and colour.

In terms of localizing the translation, the minimalism of the set helped to place the scene in a certain place by the simple use of lighting. In the scenes taking place in the Sanatorium, the lighting creates the effect of bars on the floor. In Act II, block lighting frames spaces such as Lucy's coffin and Mina's bed.

(http://dancetabs.com/2014/09/northern-ballet-dracula-leeds/)
Figure 7.9: The Sanatorium.

The costume design suggested the suppression and morality specific to the Victorian society. The contrast between masculinity and femininity, between humankind and vampires, was reinforced through costumes. Dracula's costumes were tailored from heavy fabrics, such as velvet, to highlight his status. His large black cloak was a part of the choreography and created the effect of wings as he turned and sliced through the air. Lucy's initial costume changed from the innocent pink to ever deeper tones of red to suggest her altered state and her slipping under Dracula's spell.

(http://dancetabs.com/2014/09/northern-ballet-dracula-leeds/)
Figure 7.10: Dracula's costume.

The music is a compilation realized by Mikhail Popov. The score includes movements, among others, from Alfred Schnittke's *Faust Cantata, Concerto Grosso no. 1, (K) ein Sommernachtstraum*, from Arvo Pärt's *Spiegel im Spiegel*, and from Michael Daugherty's *Red Cape Tango* (the 5th movement of Metropolis Symphony). Schnittke's compositions juxtapose a variety of styles, creating a mix of tonal and melodic compositions in a method known as polystylism.

The music of the ballet proposes an obvious melody and then distorts it to reflect the twisted world of Dracula. The heterogeneity of the general score is counterbalanced by some effects that work like refrains. The sonority of bells and solo violin pieces permeate the major pieces of music in Dracula and they support the perpetuation of an overarching theme.

The choreography displays various rhythms in accordance with the intensity of the events. In Act I, while lawyer Jonathan gazes at his beloved Mina's photograph, he gets invaded by the Brides of Dracula, a situation that triggers a stormy, sensuous and lethal dance. Another dynamic movement is created by Dracula skittering among his fellow bats. People hunting Dracula and the post-mortem vampire Lucy at night form a gruesome and spectacular scene.

(https://www.pinterest.com/pin/509610514058581494/)

Figure 7.11: The Brides of Dracula.

As we can see, this intersemiotic translation of Bram Stoker's *Dracula* created a much more spectacular and aesthetically valid version than the novel itself. There are other ballet versions of famous literary works. Northern Ballet, for instance, staged *The Great Gatsby, Peter Pan, Hamlet* and *Beauty & the Beast*. Not only did many of these transmutations create exquisite shows, but they proved to be resourceful in terms of reinventing and advertising – through their syncretism – notorious, but less and less frequented artistic pieces.

7.6 Fingerspelling and Non-Verbal Communication

Another type of intersemiotic translation is the use of the American Sign Language (ASL). In this type of language, signs are produced by moving the hands in combination with facial expressions and postures of the body. This is the primary language of deaf or hard-of-hearing people. This is not a universal language, as the signs used are different in various countries. For example, British Sign Language (BSL) contains other signs than ASL and is reported to be less rich in 'alphabet' signs.

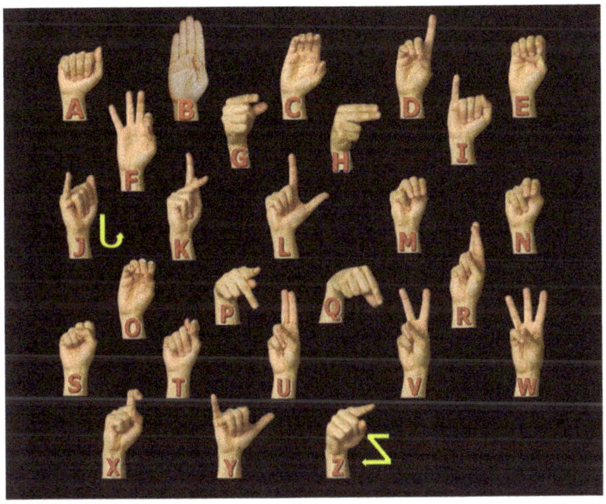

(http://www.lifeprint.com/asl101/fingerspelling/fingerspelling.htm)
Figure 7.12: American Sign Language (ASL).

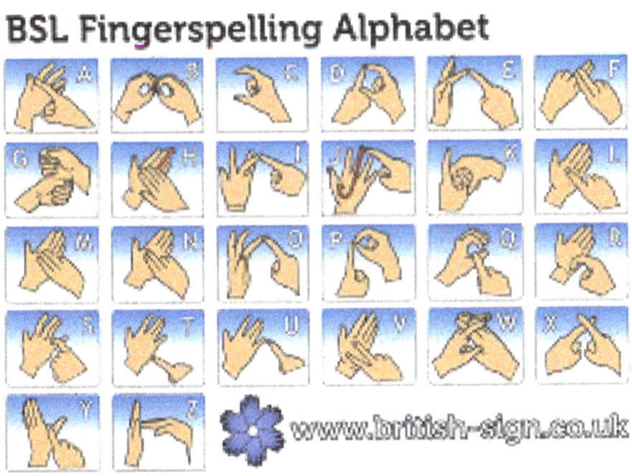

(http://www.british-sign.co.uk/british-sign-language-bsl/free-fingerspelling-chart/)
Figure 7.13: British Sign Language (BSL).

The beginnings of ASL are cloudy, but it is known that more than 200 years ago it arose from the intermixing of local sign languages and French Sign Language (LSF, or Langue des Signes Française).

In spite of containing all the fundamental features of a language (its own rules for pronunciation, word order, and structured grammar), ASL is a language completely distinct from English. One could identify similarities in the conversational behaviour: if English speakers ask a question by raising the pitch of their voice, ASL users raise their eyebrows, widen their eyes and tilt their bodies forward in the same situation.

Just as any other language, ASL has regional versions and dialects, but also regional variations in the rhythm of signing, form, and pronunciation.

One salient difference between an ASL system and BSL one is that the latter uses two hands to interpret each letter instead of ASL's one. By using two hands, the degree of translatability of signs increases as this makes the hand sign look more like the letter it is representing.

But, BSL has a more intricate way of representing gender, since mom/dad/sister/brother signs are associated with different parts of the body, whereas ASL signs for females are mostly around the mouth and those for males are around the forehead area.

How relevant and necessary is this sign language or fingerspelling? Suffice it to say that there are job openings for sign language interpreters. Such interpreters are used in schools, hospitals or government agencies. One of the requirements is to possess strong skills in English and communication. The U.S. Bureau of Labour Statistics (BLS) reported that in May 2013, the median yearly salary for interpreters specialising in sign language was $ 42,420.

7.7 Tattoos and their Overlapped Meanings

Another cultural field with intersemiotic implications is that of tattoos. There are plenty of models for them and they are supposed to be translated in accordance with rigorous descriptions. Thus, the 'strength tattoos' are used as amulets, talismans or touchstones for personal virtue. The *koi fish*, for example, swims up the waterfall to become a dragon. Star tattoos, then, conjure celebrity. The symbolism of a star depends on the number of points it has and on their orientation. Stars suggest truth, hope, whereas their nocturnal appearance represents the struggle against the forces of the unknown and of darkness. The most famous stars are the Pentagram, the Nautical Star (five-pointed), and the Hexagram or the Star of David.

The *Tribal/Maori* tattoos use Polynesian patterns derived from straight-line geometric model, but also with spirals. These tattoos were realized by cutting the skin, not only engraving it with ink, and signalled different stages of self-fulfilment, like entering the adult age. Nowadays, they are engraved only for the sake of look, in disregard with their former role and position on the body (especially, on the face). So, we could infer that their translation in the modern world mainly equates to a vulgar and a disenchanted approach.

(http://www.freetattoodesigns.org/maori-tattoos.html)
Figure 7.14: Polynesian tattoo.

Oppositely, we have Crosses symbolizing faith, hope, and commemoration of the deceased. The vertical and horizontal lines of the cross may represent Father and Mother Nature. The cross tattoos appear in various shapes with different significations: ankh or tau, swastika or Thor's Hammer, crux ansata or cross with a handle (Coptic cross), denoting power over material nature. *Ankh* tattoos are related to the ancient Egyptian symbol for life. They consist of a looped cross in which, the ellipse at the top suggests the ascendancy of spirit into the nether world over the cross of matter (this world).

(http://symboldictionary.net/?p=641)
Figure 7.15: *Ankh* tattoo.

Angel wings evoke God's protection, but when wings are associated with fairies, butterflies, dragonflies, griffins and the winged-horse Pegasus; they contain the alchemical and magical transformation through which an individual gains access to a superior position in the universal hierarchy.

There are tattoos that come more naturally to women, as it is the case with *butterflies*, which work as emblems of nobility and spiritual renaissance. The ancient Greeks also imagined Psyche (soul) with butterfly wings.

Sun tattoos make reference to the life-giving deity worshipped in the pantheon of ancient peoples. In many cultures, there are myths detailing an apocalypse during which the sun is destroyed or devoured. But, the sun rises again in a cycle of life and dark, regeneration and reincarnation.

There are also more difficult to translate symbols. *Acorn* tattoos resume ancient beliefs that this fruit warded off evil spirits. Old English folklore sustained that those women who carried acorns kept wrinkles at bay.

Alligator tattoos are manly as they invoke the supremacy on the food chain of this amphibious animal.

Ambigram tattoos are graphical figures that spell out words in a twisted manner.

(http://pixshark.com/ambigram-tattoo.htm)
Figure 7.16: *Ambigram* tattoo.

Anamorphosis tattoos are also a form of art that does not reflect hidden meanings, but creates impressive visual effects. Such a technique generates optical illusions able to trick the eye.

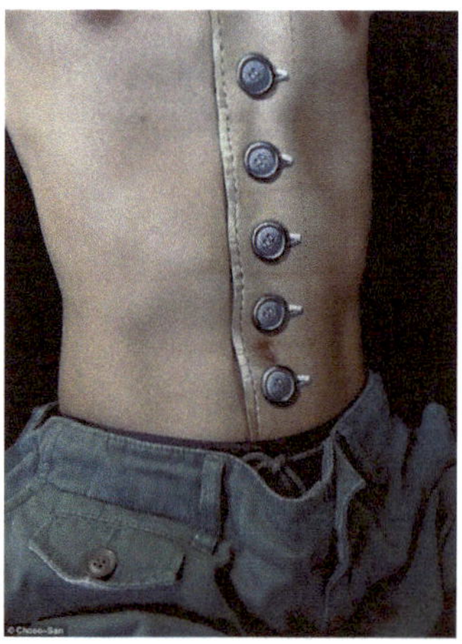

(http://www.viralsaurus.com/2014/02/girl-leg-fascinating-latest-craze-body-art/)
Figure 7.17: *Anamorphosis* tattoo.

There may be pictures that conceal messages or images within a design. The hidden elements can be distinguished only if the observer shifts position and finds the proper perspective. In these cases, translation is the equivalent of a discovery and it necessitates investigating abilities. But, there may be also illusionary effects created by a 3D perspective, as in a text illustrated by life-like reproductions.

Anemone Flower tattoos are associated with death and sleep. In the myth of Aphrodite mourning the death of Adonis, the flower sprung from her tears and that is why they are preferred by women.

(http://fc01.deviantart.net/fs71/i/2012/052/a/9/stargazer_lilly__s_and_anemone_flowers_by_phantomphreaq-d4qip3j.jpg)

Figure 7.18: *Anemone Flower* tattoo.

Among the tattoos specific to one of the genders or attached to some parts of the body, we have *ankle* tattoos. A tattoo of this kind is the apple, a fruit having long symbolized fertility, love, sensuality, sin and temptation. But also, it may recall the promise of sweetness as it turns from green to red in the rays of the sun.

Barcode tattoos are meant to be ironical and to warn about the extinction of culture and people becoming serialized products.

There are many tattoos related to the Buddhist faith and maybe, they represent the most complex category in this field. Their initial religious message has slipped towards an all-encompassing spiritual perspective, more suitable to a postindustrial, entertainment based society.

Buddha's eyes tattoos look like a pair of giant eyes symbolizing Buddha's omniscience and encouraging compassion towards all living creatures.

(http://college.holycross.edu/projects/himalayan_cultures/2011_plans/dmackiew/)
Figure 7.19: *Buddha's eyes* tattoo.

Buddha's Footprint indicates that the god was initially a mortal who walked the earth. We are reminded by this imprint that our spiritual life should conserve the contact with the material plan of existence so that, it would not fall into the trap of loftiness.

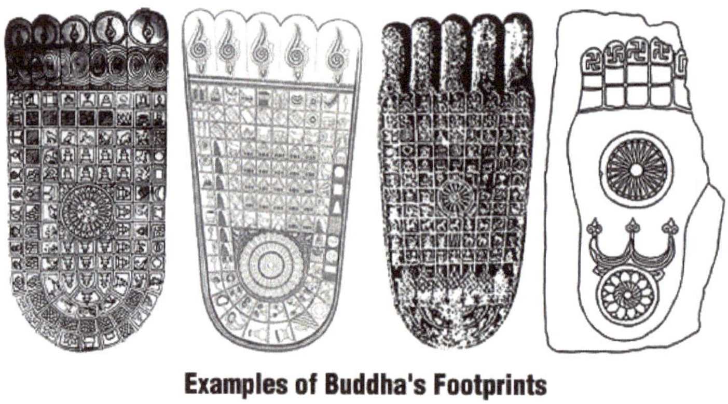

(http://www.vanishingtattoo.com/tattoos_designs_symbols_buddhas_footprint.htm)
Figure 7.20: *Buddha's Footprint* tattoo.

The Buddhist Golden Fish tattoos – usually figures as a pair – symbolize the state of temerity while afloat on a sea of suffering (the sea of *samsara*). The golden fish persisted as a sign of happiness and emancipation, the result of the liberation experienced when the initiate takes Buddha's teachings to heart.

(http://www.vanishingtattoo.com/tds/tattoos_designs_symbols_buddha_golden_fishes_inspiration.htm)

Figure 7.21: *Buddhist Golden Fish* tattoo.

The *Buddhist Knot* tattoos, or the *Endless Knot*, resemble the Celtic Knot, and look like an unbroken weaving of geometric lines.

(http://www.vanishingtattoo.com/tds/tattoos_designs_symbols_buddha_knot_inspiration.htm)

Figure 7.22: *Buddhist Knot* tattoo.

This knot is one of the *Eight Auspicious Symbols* in Buddhism and is referred to as the 'Mystic Dragon'. It also symbolizes Buddha's infinite wisdom and compassion for every being.

Buddhist Victory Banner tattoo, another of the Eight Auspicious Symbols of this spiritual movement, signifies Buddha's enlightenment through his conquering of lust, pride, passion and fear of death. These are considered the four pitfalls or 'Maras', which hinder the final liberation, Nirvana.

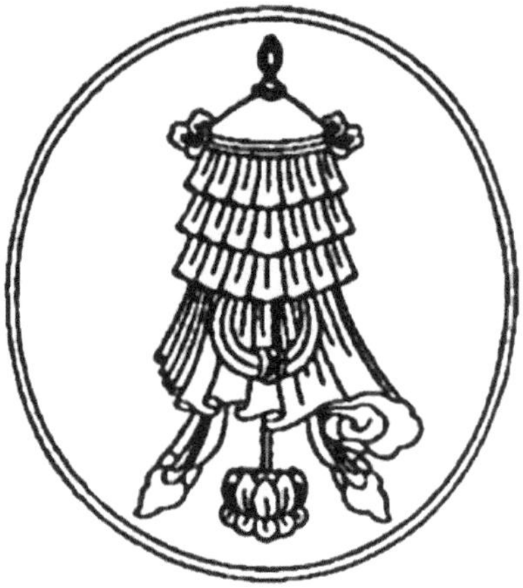

(http://orgs.usd.edu/nmm/Tibet/Symbols/TibetanAuspiciousSymbol7VictoryBanner.jpg)
Figure 7.23: *Buddhist Victory Banner* tattoo.

Eye of Horus tattoos are the symbol of the Egyptian Falcon Sky God. The eye was a symbol of indestructible royal power and rebirth. Freemasonry also adopted this symbol as the Eye of Providence, and in this capacity, it appears on the recto of the Great Seal of the United States.

(http://www.tiptopsigns.com/Eye-of-Horus-Egyptian-Decal-Sticker-p-4788.html)
Figure 7.24: *Eye of Horus* tattoo.

There are, then, *Family Crest* tattoos:

(http://www.redlegger.com/apps/photos/photo?photoid=102726170)

Figure 7.25: *Family Crest* tattoos.

Coat of Arms tattoos:

(https://www.pinterest.com/pin/493918284106358218/)

Figure 7.26: *Coat of Arms* tattoos.

Clan tattoos and *Heraldry* tattoos for celebrating family ties and heritage. The message to be translated from their representations is respectability, affiliation and historicity in terms of lineage.

(http://galleryhip.com/scottish-clan-tattoos.html)
Figure 7.27: *Clan* tattoo.

Flaming tattoos are the status symbol for trailer park residents throughout America. The flamingo is the unofficial landmark of Florida and of travel to tropical destinations and leisure.

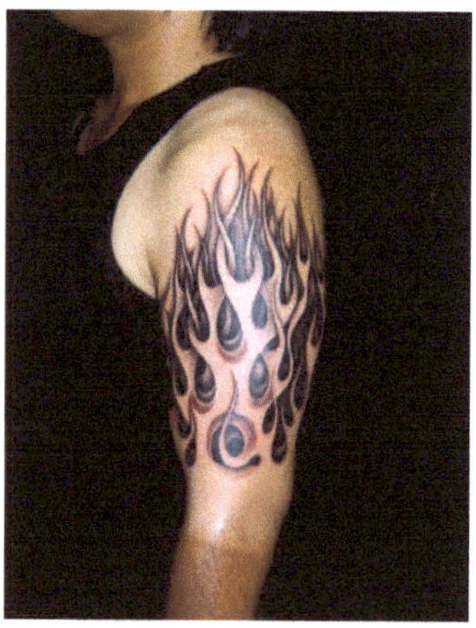

(http://www.tattoos20.com/category/fire-flame-tattoos/)
Figure 7.28: *Flaming* tattoo.

Ganesh tattoos represent the deity with the head of an elephant and the body of a human. Ganesh was the Asian Lord of Success and Destroyer of obstacles, but also the god of Everyman. Salman Rushdie, in The Satanic Verses, plays upon the symbolism of this deity by basing the sexual success of a Bollywood actor on his constant acting as Ganesh in Indian cinematographic super-productions. All the women who make love to him do not want to see his real visage, so he is forced to put on the elephant-like mask of Ganesh even during intimate moments. The trunk also has the role of enflaming the sexual symbolism.

(https://www.pinterest.com/pin/558446422519283472/)
Figure 7.29: *Ganesh* tattoo.

Closer to the 'classical' semiosis are the *alphabet* tattoos, wherein letters represent various aspirations or reminders. All in all, in our globalized and corporatist world, tattoos represent more often than not an instance of ignorance or incomplete knowledge, at least. This practice is a case in point of mistranslation of ancient symbols within the context of a linguistically and anthropologically impoverished world.

7.8 Conclusions

The transfer of signs between heterogeneous systems respects, in a certain degree, the algorithm of the ATRIA model as it implies launching themes with known information (IT) and interpreting them with the help of rhemes, which carry new information (THERE). The themes of the text make up a holon, i.e., 'a system which is a whole in itself as well as part of a larger system' (Dejica, 2008). Such holons can be specific to certain cultural systems of the source or target languages. It results that the various methods of translation are perfectly applicable in the case of incongruous systems of signs. Whether we translate texts into other texts or images into texts, sounds into images, films into texts and so on, the principles of translation are the same. Intersemiotic translations can be simple or complex – depending on the difficulty of the process of decoding the messages involved, but the expertise of translators must be, in these situations, more comprehensive, as the range of signs is broader. The conclusion is that all translators should be open-minded and ready to accept new professional challenges.

References

Dejica, D. (2008). Using Hol-Atomistic and Holistic Lexical and Grammatical Relations in Translation, In Superceanu, D. & D. Dejica (eds.), Professional Communication and Translation Studies, Volume 1, Issue 1-2, Timişoara: Politehnica University Press, 147-151.

Eco, U. (2001). Experiences in Translation. Trans. by A. McEwen. Toronto, Buffalo, London: University of Toronto Press, 67.

Gambier; Y. & Gottlieb H. (Eds.) (2001). (Multi) Media Translation: Concepts, Practices, and Research. Amsterdam/Philadelphia: John Benjamins Publishing Company, p. XI.

Greimas, A.-J. (1966). Sémantique structurale. Recherche de méthode. Paris: Larousse, 12.

Hatim, B. & Munday, J. (2004). Translation. An Advanced Resource Book. New York & London: Routledge, 5.

Hervey, S., Higgins, J., Cragie, S., & Gambarotta, P. (2000). Thinking Italian Translation. A Course in Translation Method: Italian to English. New York & London: Routledge, 9.

Jakobson. R. (1959). On linguistic aspects of translation [2004], in L. Venuti (Ed.) (2004). The Translation Studies Reader. New York & London: Routledge, 138-143

Kress, G. & Leeuwen, T. V. (2001). Multimodal Discourse. The Modes and Media of Contemporary Communication. London: Hodder Education, 1-4.

Kuhiwczak, P. & Littau, K. (Eds.) (2007). A Companion to Translation Studies. Clevedon: Multilingual Matters Ltd, 4.

Landers, E. C. (2001). Literary Translation: A Practical Guide. Clevedon: Multilingual Matters Ltd, 7.

Lefevere, A. (Ed.) (1992). Translation/History/Culture. A Source Book. London & New York: Routledge.

Martin, B. & Ringham, F. (2006). Key Terms in Semiotics. London & New York: Continuum International Publishing Group, 108.

Metcalf, A. (2002). Predicting New Words. The Secret of Their Success. Houghton Mifflin Company, Boston New York.

Mitchell W.-J.-T. (1994). Picture Theory: Essays on Visual and Verbal Representation. Chicago & London: University of Chicago Press.

Petrilli, S. (2003). Translation and Semiosis. Introduction. In S. Petrilli (Ed.), Translation, Translation (pp. 17-37). Amsterdam & New York: Rodopi, p. 31.

Snell-Hornby, M. (2006). The Turns of Translation Studies. New Paradigms or Shifting Viewpoints? Amsterdam/Philadelphia: John Benjamins Publishing Company.

Torop, P. (2004). Locations in Intersemiotic Space. In V. Sarapik (Ed.), Place and Location: Studies in Environmental Aesthetics and Semiotics IV (pp. 59-68). Tallinn: Estonian Academy of Arts, p. 64.

Venuti, L. (2004). The Translator's Invisibility. A History of Translation. London & New York, Routledge, p. 5.

William, J. & Chesterman, A. (2002). The Map: A Beginner's Guide to Doing Research in Translation Studies. United Kingdom, Manchester: Jerome Publishing, 7.

Webography

http://study.com/articles/Sign_Language_Interpreter_Job_Description_Duties_and_Requirements.html, accessed 19 September 2015

http://study.com/articles/Sign_Language_Interpreter_Job_Description_Duties_and_Requirements.html, accessed 19 September 2015

http://www.nidcd.nih.gov/health/hearing/pages/asl.aspx, accessed 19 September 2015

http://sparkleteaching.blogspot.ro/2011/05/free-form-poem-mouses-tale.html, accessed 15 September 2015

Titela Vîlceanu

8 Evaluating Online Resources for Terminology Management in Legal Translation

8.1 Setting the Scene. The Romanian Legal Framework and Legal Translation Market

The Romanian legal translation market has evolved steadily, expanding after Romania's accession to the EU in 2007. We acknowledge the quantitative development as demand-driven (translation means service provision, it is commissioned, it does not take place in a social vacuum) as well as quality increase, although the two components show a significantly different degree. Adopting an in-depth approach, we may identify several factors that have contributed heavily to the shaping of the Romanian legal translation market, such as follows:

- the need for legal translation – we use the term in its broadest sense, encompassing the translation of all kinds of legal documents, but laying special emphasis on the EU law
- the time pressure (enforced deadlines)
- the insufficient number of legal translation training programmes: at the university level, Bachelor's programmes of the major higher education institutions across Romania include a relatively small number of specialised translation and terminology courses – business translation, medical translation, technical translation, audio-visual translation, general terminology course, etc. The situation is replicated at the level of the Master's programmes, where a module in legal translation comes under the umbrella of specialised translation, and where legal terminology is not in focus – there is one notable exception, i.e., Master's programme in *English and French Languages. European Legal Translation and Terminology*, University of Craiova, where translator training underpins development of legal translation skills, legal terminology management skills, on a par with domain specialisation (one-third of the courses is dedicated to familiarisation with different branches of European law and with European bodies) (source: http://cis01.central.ucv.ro/litere/oferta-educationala/GhidTTJE2013.pdf).

In spite of the reform of the higher education system (through the implementation of the Bologna process), of the change to competence-based curricula and the constant endeavour of universities as providers of translator training programmes to map the labour market and achieve harmonised translator's multilayered competence at the European level, having the EMT flagship model at the core (please visit http://ec.europa.eu/dgs/translation/programmes/emt/key_documents/emt_competences_

translators_en.pdf) – for instance, 5 Master's programmes in translation delivered by 5 major Romanian universities (more precisely, University of Bucharest – *Masteratul pentru traducerea textului literar contemporan / Master's in Contemporary Literary Translation*; University of Cluj-Napoca – *Masterat european de traductologie – terminologie / European Master's in Translation Studies – Terminology*; University of Craiova – *Limba engleză și limba franceză. traducere și terminologie juridică europeană / English and French Languages. European Legal Translation and Terminology*; University Dunărea de Jos of Galați – *Traducere și interpretare / Translation and Interpretation*; University Transilvania of Brasov – *Traducere și interpretariat din limba franceză în limba română / Translation and Interpretation from French into Romanian*) belong to the OPTIMALE (*Optimising Professional Translator Training in a Multilingual Europe*) ERASMUS Academic Network, which comprises 70 partners from 32 different European countries (out of which 27 are within the EU), and whose mission statement reads 'aims to act as a vehicle and stimulus for innovation and high quality in the training of professional translators' (source: www.translator-training.eu/, http://www.ressources.univ-rennes2.fr/service-relations-internationales/optimale/map/) – there is still a divide between the universities and the (legal) translation market. Law 178/1997 with the subsequent modifications and additions (still in force, although a new law has been submitted to public debate) provides that the legal translator may be authorised by the Ministry of Justice, based on adequate qualifications. The question arises: *What do these qualifications or eligibility conditions refer to*? The answer points out to what we would label as *incongruity* between the above mentioned aforementioned professed values and realities of the legally recognised profession: a Bachelor's Degree in Foreign Languages or a Baccalaureate Diploma (A-level) in the case of graduates from an international high school (foreign language-taught programme) or a Translator's Certificate in the field of law, issued by the Ministry of Culture (source: websites of the Ministry of Justice, http://www.just.ro/Sectiuni/Informatiiutile/interpretisitraducatori_22022013/tabid/2422/Default.aspx, and the Professional Association of Translators and Interpreters – APIT (Romanian acronym), www.apit.ro/informatii-utile.html).

Admittedly, legal translators are authorised following no open competition, and the ever increasing market and societal demand and offer of highly qualified translators seems to be reduced to the natural selection or 'survival of the fittest' principle. At what costs? To our best knowledge, low professional relevance or visibility, a large number of poor quality translations, diminished translator's fees and turnover (also due to the large number of authorised legal translators acting as freelance or in-house ones or cumulatively), and reduced motivation for pursuing professional development come top-most and should spread wider and deeper concerns among policy makers and management authorities.

Needless to say those current and emergent competence requirements at the European level – which, once again, universities seek to meet via competence-based curricula, work placements and employers' consultation – are disregarded in the legal provisions in force. We reinforce the idea by mentioning that within the OPTIMALE framework, the online survey jointly conducted with the EUATC (European Union of Associations of Translation Companies), involving 680 employers from across Europe, revealed that 'A university degree in translation or related areas (and not simply a modern language degree)' is a prerequisite in the translation industry (source: http://www.ressources.univ-rennes2.fr/service-relations-internationales/optimale/attachments/article/40/Public%20part_report_2010_OPTIMALE%204018-001-001.pdf).

8.2 Thinking Legal Translation

8.2.1 Competence-related Considerations

The legal translator's multilayered competence underpins both a process- and product-oriented approach, although it seems that the product dimension has taken precedence, being the most visible part of translation as design-oriented, precise and measurable (complying with the specifications). A professional translator should master the methodological toolkit, conceptual frame and related terminology. The legal translator's accountability is expressed in terms of cost-effectiveness (efficiency) and effectiveness. Moreover, it should be understood that effectiveness and efficiency derive not only from an empirical approach, but they are also based on proactive behaviour, i.e., the translator's seeking to anticipate problems and identify re-usable or generalisable solutions. Therefore, the inward-looking perspective of the translator should be complemented by the outward-looking one (against a set of objective criteria).

One important component of the legal translational know-how in relation to the selection and use of available resources is *information mining competence* (in accordance with the EMT framework), broadly defined as the ability to retrieve and evaluate information in a multiplicity of formats and for a multiplicity of purposes. Legal translators should be able to decide what kind of information they need to gather in order to create a context for the source language text and to fully integrate the target language text into its situational context (ultimately reduced to *skopos* as translators should master top-down and bottom-up information processing strategies, adopting an action-oriented approach). They should be able to locate primary and secondary information sources by tracing available resources and eliminating unnecessary or outdated sources. There is a wide range of resource types that translators should be aware of (dictionaries, glossaries, catalogues, indexes, workbenches, parallel texts, etc.) in order to extract and manage relevant information.

Effective and efficient legal translators will become translation memory managers in the sense that they will be able to use translation memory software, storing translated texts as reference materials, creating databases with different task-related information sources, thus accelerating the translation process and fostering professional development (long-term orientation).

Another component to be activated in the evaluation of (online) resources for legal translation is represented by *thematic area competence*, i.e., sufficient field knowledge for functional adequacy. Accordingly, legal translators should develop discovery skills and search for information related not only to the topic area of the document in question (using keywords in the text), but also expand the search to detect the hierarchical structure of themes (in the form of thematic maps). The next step is to acquire knowledge of the general terminology in the field and general stylistic features of the document type, appply such knowledge deductively and/or by association, and further detect specificities.

Thematic analysis underlies an exploratory spirit, mental alertness to the myriad of relations between different themes that make a corpus consistent and intelligible.

Cumulatively, the two components, which we see to be interrelated, lend priority to observation over intuition, to process orientation over product orientation, integrating quantitative analysis to qualitative statements about the usability of resources, language use, terminology management and legal cultures.

Nevertheless, we should not become too optimistic or idealistic. Terminology mining is time-consuming (Picht in Austermühl, 2001: 102, endorses that it takes up 75% of the translation time) and in practice, more often than not, it happens that translators do not have the time to evaluate all the available resources, thus, the selection of reliable tools may be done based on prior successful experience, immediate availability, peer advice, etc. Besides, the dynamics of the European law making is another barrier in the translator's coping with the huge bulk of texts, evolution of legal concepts, detection of the national or supranational character of the legal term in question, etc.

8.2.2 Overview of Online Resources

8.2.2.1 Objectives

Our main aim is to raise critical awareness of the diversity, reliability, and usefulness of the information provided by the different types of online resources from a translation-oriented perspective: bilingual legal dictionaries, multilingual legal glossaries (term bases), parallel corpora and interactive tools (discussion forums).

Secondly, we intend to raise awareness of the translators' research potential with a view to informed decision-making and closer mapping of theory to practice.

Thirdly, we aim to provide a toolkit to increase the legal translators' efficiency and effectiveness in the management of terminology.

8.2.2.2 Methodology

We shall use a mixed research methodology, combining both quantitative methods – collection of empirical data, statistics, and qualitative ones – review of mainstream literature, analysis of the reliability and usefulness of online resources, identification of patterns and relationships between online resources.

Our approach is mainly descriptive, yet, we would like the final recommendations to be considered from a norm-oriented perspective as guiding legal translators' behaviour due to their evidence-based nature. In this respect, the overview of online resources shall focus on the general features understood as strategic costs and benefits.

In the applied part, for the purpose of the current paper and due to space constraints, we shall restrict our investigation to two legal terms, i.e., *court* and *tribunal*, which may be considered partial synonyms, but which, nevertheless, are not freely interchangeable and describe different legal systems.

8.2.2.3 The burden of proof

Bilingual legal dictionaries: English–Romanian, Romanian–English

Generally speaking, bilingual legal dictionaries provide decontextualised correspondents or standard equivalents, with little detailed information on typical collocations, legal areas, contexts of use, examples and differences between the source language concept and the target language counterpart. Even if equivalents provide a viable solution, their cope of application is not the same. At this point, we make the terminological distinction between *correspondent* and *equivalent*, the former being decontextualised and predefined, whereas the latter involves a context-embedded decision-making process and a gradient – there are no absolute or perfect equivalents, they may range from *non-equivalents, partial equivalents* (minimum degree) to *near* or *optimal equivalents* (maximum degree) (see Catford, 1965; Newmark, 1988; Sandrini, 1999; Šarčević, 2000; Groot & Laer, 2006, etc.).

Another shortcoming is related to polysemous items that may become a stumbling block for translators in the attempt to disambiguate. As noted by different scholars, bilingual dictionaries of law are even 'less informative than monolingual dictionaries' (Šarčević, 1989: 277). 'The dubious quality of legal dictionaries' (Groot & Laer, 2006) is assessed based on the system specificity of legal terms and on the need to frequently revise and update them.

A search of *English–Romanian/Romanian–English legal dictionary* detected a general dictionary incorporating legal entries (http://dictionare.4rev.net/?Get=tra&Dictionary_base=0&Dictionary_list) and EUdict (http://www.eudict.com/?lang=engrom&word=court).

Entry: *court*

In the first dictionary, section *Standard dictionary, English-Romanian*, we see that equivalents are provided at the level of the phrase. Comment: as far as the polysemy of *court* is concerned, its different meanings are listed alphabetically (with no specification of domain). The legal meaning is exemplified as follows:

admiralty court – tribunal maritim
court of arbitration – arbitraj (tribunal, curte)
court martial – curte marțială
court of justice – tribunal
law-courts – tribunal
main court – curte de onoare
maritime court – tribunal maritim
police-court – tribunal corecțional

Entry: *tribunal*. No results found.
If we undertake validation work (to check consistency or back-translation), in the section *Romanian-English*, under the headword *curte*, we simply find the enumeration of the items *court, yard, courtyard* and *tribunal*. With reference to the equivalents of the Romanian *tribunal*, these are provided phraseologically, too:

tribunal, court; <~militar> military tribunal; <~militar> Court Martial; <~popular> People's Court; <~suprem> Supreme Court.

Comment: in subsidiary, an item belonging to the lexical field of *tribunal* is listed: *ușier (la tribunal) - bailiff, usher of the court*. Furthermore, we shall see, after consultation of glossaries (IATE), parallel corpora (EurLex) and interactive tools (Proz.com), that the equivalence of *bailiff* to *ușier* is referentially inaccurate.

EUdict - English-Romanian section:
(the) Law Courts; <~de ocol/pace aprox.>country court - judecătorie (în Londra)
1. jur. to rule smb. out of court. 2. fig. to exonerate smb. (from blame) - a scoate pe cineva din cauză
country court – judecătorie (mai mică)
court-martialled – tradus în fața curții marțiale
in court – la judecată
in the first court/instance – în prima instanță
jur. to discharge an order of the court – a revoca o hotărâre a curții judecătorești
to lay a matter before the court – a ridica o chestiune în fața tribunalului
law court – judecătorie (sediu)
court martial – curte marțială

EUdict English-Romanian Dictionary:
curte – tribunal
tribunal

a fi citat la tribunal – to come up before the bench
ușier (la tribunal) – bailiff
ușier (la tribunal) – usher to the court

Comment: both noun phrases and verb phrases are taken into consideration.
Note: In the case of EUdict, there is a disclaimer: 'EUdict (European dictionary) is a collection of online dictionaries for the languages spoken mostly in Europe (414 language pairs). These dictionaries are the result of the work of many authors who worked very hard and finally offered their product free of charge on the internet thus making it easier to all of us to communicate with each other. Some of the dictionaries have only a few thousand words, others have more than 250,000. Some of the words may be incorrectly translated or mistyped'. However, it also aims, rhetorically, to create a feeling of togetherness, of a community of practice, and it counts as a covert invitation addressed to those valuing the 'hard work' to contribute to this tool development.

Multilingual legal glossaries
Terms should be envisaged as depositories of structured knowledge in a particular field. With strict reference to legal terms, their meaning may be shaped and stabilised intertextually (by legislation); therefore, when two legal systems are involved, the source language term and its equivalent in the target language will have a different semantic potential (for instance, narrowing or extension of meaning). Moreover, the equivalent should provide information about the legal system as a knowledge base. Equivalents have their own lifecycle: they are created, adopted by a speech community or a community of practice and if used repeatedly, they entrench cognitive routines, becoming *established* (Molina & Hurtado Albir, 2002: 510).
IATE – the EU's multilingual term base, interactive terminology for Europe (source: http://iate.europa.eu/) is a web-based tool developed by language services with the main aim of benefitting institutional translators and language staff in their tasks. IATE is inter-institutional in nature, as indicated by the fact that the entries specify the institution from which an input is retrieved. Besides, the reliability of each term and the date of registration are also made available, on a par with the definition of the original term. On the other hand, definitions of the equivalent term and illustrations of use in the target language are not common practice.
Focussing on the analysis of the different renderings of the term *court*, we identified 72 entries, pertaining to different domains and sub-domains. The following examples are illustrative:

LAW, EU institution [Council]
Rules of procedure of the General Court (rated by 3 - reliable). Term reference: *Consolidated version of the Rules of Procedure of the General Court* of 2 May 1991, *OJ C* 177, 2.7.2010, p. 37, 32010Q0702(02). Date: 16/07/2010 - *Regulamentul de procedură*

al Tribunalului (rated by 3), *Versiunea consolidată a Regulamentului de procedură al Tribunalului din 2 mai 1991, Jurnalul Oficial* C 177, 02/07/2010, p. 0037 - 0070, 32010Q0702(02)/RO. Date: 03/08/2010.

LAW, EU body, Justice, EU institution [Council]
Court of Justice / European Court of Justice (rated by 4 - very reliable). Term reference: *Consolidated version of the Treaty on European Union, Article 19(1)*,http://eur-lex.europa.eu/Lex... and *Consolidated version of the Treaty on the Functioning of the European Union*, Articles 251 to 253, http:/eur-lex.europa.eu/Lex.., 15/05/2014 – *Curtea de Justiție* (rated by 4). Term reference: Articolul 19 alineatul (1) din *Versiunea consolidată a Tratatului privind Uniunea Europeană*, http://eur-lex.europa.eu/Lex. Date: 15/05/2014.

Note: the former English phrase (*Court of Justice*) is marked as preferred.

European construction, EU institution [Council]
Court of Justice of the European Union (rated by 4 - very reliable). Term reference: *Consolidated version of the Treaty on European Union*, http://eur-lex.europa.eu/Lex... Date: 22/09/2014 - *Curtea de Justiție a Uniunii Europene* (rating - 4). Term reference: Articolul 19 alineatul (1) din *Versiunea consolidată a Tratatului privind Uniunea Europeană*, http://eur-lex.europa.eu/Lex. Date: 20/05/2014

Note: in this case, the entry date of the Romanian equivalent term precedes that of the English one, which means that the latter was updated.

Administrative law, LAW [EP]
bailiff / court officer (rated 3). Term reference: JA 96/636/JHA (1) OJ L 287/96, p. 3 Dir. 92/51/EEC (2) OJ L 209/92 p. 43. Date: 03/05/2002 - *executor judecătoresc* (rated 3). Term reference: *Ghid pentru traducerea în limba română a jurisprudenţei Curţii Europene a Drepturilor Omului*, 2010, www.ier.ro.

Note: the Romanian equivalent is recorded eight years later, which may be an indication of the fact that it gained relevance then.

LAW [COM]
court order/ court ruling / judicial decision (all of them rated 2 - minimum reliability). Term reference: *Judicial decision: Proposal for a Directive of the European Parliament and of the Council on measures and procedures to ensure the enforcement of intellectual property rights* (COM(2003) 46 final). Date: 13/02/1991 - *hotărâre judecătorească* (rated 3). Term reference: *Guvernul României Hotărârea 210 din 27 februarie 2008 privind aprobarea prealabilă, în vederea recunoașterii prin hotărâre judecătorească, a*

Fundației *'Kinderen in Nood' din Olanda*, Publicat în *Monitorul Oficial* 170 din 5 martie 2008, http://legestart.ro/Hotarare...Date: 01/03/2012.

Tribunal - 9 entries
Legal system, Public international law [Council]
International Tribunal for the Law of the Sea (rated 4). Term reference: website of the International Tribunal for the Law of the Sea, http://www.itlos.org/. Date: 15/05/2014 - *Tribunalul Internațional pentru Dreptul Mării* (rated 4). Term reference: Centrul de Informare ONU pentru România, *Convenția Națiunilor Unite asupra dreptului mării*, http://www.onuinfo.ro/docume...[7.2.2011]. Date: 10/04/2014.

Legal system, EU institution [Council]
Civil Service Tribunal / European Union Civil Service Tribunal (both rated 4). Term reference: *Annex I to Protocol (No 3) on the Statute of the Court of Justice of the European Union, Consolidated versions of the Treaty on European Union and the Treaty on the Functioning of the European Union*, http://eur-lex.europa.eu/Lex...Date: 20/05/2014 - *Tribunalul Funcției Publice a Uniunii Europene* (rated 4). Term reference: *Versiunea consolidată a Tratatului privind Uniunea Europeană și a Tratatului privind funcționarea Uniunii Europene, Protocolul nr. 3 privind statutul Curții de Justiție a Uniunii Europene, anexa I*, http://eur-lex.europa.eu/Lex...Date: 20/05/2014.

Note: when the information is extracted from Eurlex, we notice the simultaneous recording of the source language term and target language term.

Parallel corpora

Parallel corpora (also known as translation corpora) are identified as source language texts aligned with the corresponding target language texts or translations. We focus on the EU law as overrepresented in online translation resources in comparison with other legal genres; therefore, in line with Kasirer (2001), we advocate *legicentrism*.

There are skeptical voices claiming that such corpora are of little assistance to achieve a high degree of 'descriptive adequacy' (Granger, 2003: 19) and that they have a limited applicability to forensic linguists and consequently, to legal translators, by virtue of the fact that legislation 'does not often need large corpora since, because of its conservatism and "formulaic form-function correlations", it may be "equally efficient and reliable" to conduct a manual analysis, for example on a single legislative act' (Bhatia, Langton & Lung, 2004: 207). They go further as to say that 'there is very little need for comprehensive or automatic linguistic frequency measures, as they are easily identifiable manually' (2004: 212). To our mind, it may be the case with linguists and other language researchers (translation theorists, included), but it is unlikely that the industry will ever embark on such a task. Nevertheless, the authors admit that these corpora are most useful in 'researching intertextuality within and across a particular genre' (*ibidem*) – we infer that they may be explored and exploited to assimilate the characteristic features of legal sub-genres. What the authors overlook is qualitative analysis, which is able to provide data about institutional, social and cognitive factors, sharpening the legal translator's awareness of similarities and differences.

We acknowledge the existence of several reliable and useful parallel corpora provided by the EU, briefly described below, yet, we believe that EurLex deserves special attention as easily available, expanding in real time with the EU law text production and highly reliable for the validation of online dictionaries and glossaries search results.

The JRC-Acquis (https://ec.europa.eu/jrc/en/language-technologies/jrc-acquis), freely available to be used and distributed for research purposes, claims to be the biggest parallel corpus (full-text documents, paragraph aligned) in existence, comprising selected legislative texts from 1950s up to date; as far as the pair English–Romanian is concerned, the revised corpus amounted to 19,211 texts in 2009, and the Romanian texts were processed by the Romanian Academy of Sciences, which makes them an authoritative source.

The DGT-Acquis (https://ec.europa.eu/jrc/en/language-technologies/dgt-acquis), the exclusive property of the EU, a multilingual parallel corpora (full-text documents, paragraph-aligned) extracted from the *Official Journal of the European Union*, consists of documents collected from the mid-2004 to the late 2011.

DCEP - The Digital Corpus of the European Parliament (https://ec.europa.eu/jrc/en/language-technologies/dcep) comprises legislative documents (full-text documents, sentence-aligned) produced between 2001-2012. The number of English–Romanian texts amounts to 17,526.

EUR-Lex (http://eur-lex.europa.eu/), as stated on the home page, contains 'EU law and other public EU documents, *authentic electronic Official Journal of the EU* in 24 languages'. The full-texts are organised by domain, sub-domain, year of document, type of procedure, author (i.e., European body), and the type of act. There is constant concern for improvement – latest developments are announced with respect to quick search (for instance, by document number), expert search (re-grouping of fields in a more logical manner, addition of alphanumerical sorting and of the zoom functionality – the search terms are highlighted, etc.), higher visibility of national implementing measures.

With respect to our term search, *curte* is cited 68,782 times and *tribunal* registers 32,619 occurrences.

Interactive tools – discussion forums

We identified two relevant global translation communities of practice, namely *Proz.com* and *Translatorscafe.com*, where, *inter alia*, specialised translators engage in collaborative and reflective work by asking other peers terminological questions (often concerning context-dependent phraseological units) and getting their reasoned contributions so as to reduce uncertainty. One of the major strategic benefits of such forums is that they allow for an insight into what happens in the industry, while also providing solutions to recurrent terminological problems of professional translators. More experienced translators may also provide links to 'the recruitment of background knowledge' (Evans & Green, 2006: 160) and relevant legal knowledge structures, thus enhancing a more standardised translation.

We shall exemplify the proz.com search. As far as *court* is concerned, the total number of (KudoZ) results is 74. The information is not so easily retrievable as the domains are not ordered or organised in a coherent way. In the field of *Law / Patents*, it seems that the majority of queries concern the UK legal system (not our focus with legal translators' training). What is worth noticing is that the suggested equivalents are accompanied, in many cases, by explanations shedding light on the meaning of the culture-bound items. For instance, *Court of First Instance* is equated to both *Judecă*torie and *Tribunal*, and the author explains that both are concerned with 'first instance'. Some other voices seem also hesitant, while others clearly favour *Tribunal*, but the translator is finally left alone to make his/her own decision. The English term *Tribunal* (5 results) is also envisaged beyond the word level (in phrases and sentences). For instance: *the award of the arbitration <u>tribunal</u> shall be binding upon the parties - decizia <u>instanţei</u> de arbitraj va fi obligatorie pentru părţi - clauze poliţa de asigurare* . The Romanian counterpart is based on neutralisation (translating back as *instance*), and there is a specification of the sub-domain (insurance policy). More opaque culture-bound items are translated via explicitation, i.e., the abbreviation is conveyed in Romanian by the full name: *HM Courts and Tribunal Service - Serviciul Majestăţii Sale pentru curţi şi tribunale,* and clarification work is done via indication of the term reference: https://e-justice.europa.eu/content_fundamental_rights-176-sc-ro.do.

It is obvious that such an empirical approach needs to be refined and become more systematic in order to provide more viable results.

8.3 Conclusions and Recommendations

Terminological problems illustrate the need to analyse changing situational factors and macro-textual parameters for decision-making at the micro-textual level. We advocate a more global and efficient intersystemic and interlingual information retrieval and processing in legal terminological work. The acceptability of the functional equivalent term from a reader-oriented perspective becomes the keyword 'to meet formal or conceptual correspondence priorities' (Prieto Ramos, in Cheng *et al.*, 2014: 124). In other words, bilingual and multilingual resources are expected to be descriptive rather than prescriptive in nature, informing the translator's decision.

To wrap up, reliable online resources for terminology management in legal translation should provide macro-level (comparative) information about the legal systems and branches of law (text types and sub-types) in question, meso-level information about the institutionalised practices (roughly equated to standardisation work, external and internal terminological coherence), and micro-level information, such as language affinities (collocations), pragmatic contexts of use, reformulation options, etc.

Undeniably, parallel corpora provide a fresher look at lexicography, terminology and terminography since they reflect recognised translation practices (although the source language text and target language text are equally authentic from a legal point of view), containing translation units and their equivalents in the form of single and multiple units (concern for phraseology), and should be used by legal translators to complement and validate dictionaries and glossaries, especially because dedicated extraction software (CAT tools) accelerate terminological work.

Driving the agenda forward, we hope that in the near future, not only multilingual glossaries, but also online bilingual dictionaries will be based on parallel corpora to reflect the legal language and terminology in use. Last but not least, it is our hope that online terminological resources will boost the legal translators' research-driven performance for practical integration and systematisation.

References

Bhatia, Vijay K., Langton, Nicola M., Lung, Jane. (2004). "Legal discourse: Opportunities and threats for corpus linguistics". In Connor, Ulla, Upton, Thomas A. (eds.), Discourse in the Professions. Perspectives from Corpus Linguistics. Amsterdam/Philadelphia: John Benjamins, 203-231.

Catford, John Cunnison. (1965). A Linguistic Theory of Translation. Oxford: Oxford University Press.

Evans, Vyvyan, Green, Melanie. (2006). Cognitive Linguistics. An Introduction. Edinburgh: Edinburgh.

Granger, Sylviane. (2003). "The Corpus Approach: A Common Way Forward for Contrastive Linguistics and Translation Studies" in Granger, Sylviane, Lerot, Jacques, Petch-Tyson, Stephanie (eds.). Corpus-based approaches to Contrastive Linguistics and Translation Studies. Amsterdam/New York: Rodopi, 17-29.

de Groot, Gerard-René, van Laer, Conrad, J.P. (2006). "The Dubious Quality of Legal Dictionaries" in International Journal of Legal Information, vol. 34, issue no. 1, 65-86.

Kasirer, Nicholas. (2001). "François Gény 's libre recherche scientifique as a Guide for Legal Translation" in Louisiana Law Review, vol. 61, issue no. 2, 331-352.

Molina, Lucía, Hurtado Albir, Amparo. (2002). "Translation Techniques Revisited: A Dynamic and Functionalist Approach". In Meta, XLVII, vol. 4, 498-512.

Newmark, Peter. (1988). A Textbook of Translation. London: Prentice Hall International.

Picht, Heribert. (1989). Einführung in die Terminologiearbeit in Austermuhl, Frank. 2001. Amsterdam /Philadelphia: St. Jerome Publishing, p.124.

Prieto Ramos, Fernando. (2014) in Cheng, Le, Sin, King Kui, Wagner, Anne. The Ashgate Handboook of Legal Translation. Farnham: Ashgate Publishing Company, 121-134.

Sandrini, Peter. 1999. "Legal Terminology. Some Aspects for a New Methodology" in Hermes Journal of Linguistics 22, Aarhus School of Business, 101 - 112.

Šarčević, Susan. (1989). "Conceptual Dictionaries for Translation in the Field of Law" in International Journal of Lexicography, vol. 2, issue no. 4, 277-293.

Šarčević, Susan. (2000). New Approach to Legal Translation. London: Kluwer Law International.

Websites

http://cis01.central.ucv.ro/litere/oferta-educationala/GhidTTJE2013.pdf http://dictionare.4rev.net/?Get=tra&Dictionary_base=0&Dictionary_list
https://ec.europa.eu/jrc/en/language-technologies/dcep
https://ec.europa.eu/jrc/en/language-technologies/dgt-acquis
https://ec.europa.eu/jrc/en/language-technologies/jrc-acquis
http://ec.europa.eu/dgs/translation/programmes/emt/key_documents/emt_competences_translators_en.pdf
http://www.eudict.com/?lang=engrom&word=court
http://eur-lex.europa.eu/
http://iate.europa.eu/
http://www.just.ro/Sectiuni/Informatiiutile/interpretisitraducatori_22022013/tabid/2422/Default.aspx
http://www.proz.com/
http://www.ressources.univ-rennes2.fr/service-relations-internationales/optimale/map/
http://www.translatorscafe.com/cafe/default.asp
www.apit.ro/informatii-utile.html
www.ressources.univ-rennes2.fr/service-relations-internationales/optimale/attachments/article/40/Public%20part_report_2010_OPTIMALE%204018-001-001.pdf)
www.translator-training.eu/

Loredana Pungă

9 To Delete or to Add? Omissions and Additions in Two Romanian Translations of *Jack and the Beanstalk*

9.1 Introduction

The circulation of literary texts for children in languages other than those in which they were originally written has been a widely spread and very dynamic phenomenon. The reasons that have motivated translators to work so extensively and dynamically with children's literature range, as Lathey (2012) observes, are from educational to literary and commercial. Texts have been translated and retranslated to align with the pedagogical advances and requirements at various time intervals to offer the very young readership the chance to get acquainted with writings that originate in other cultures, or to adjust the language and tone of existing translations, to meet the 'demand for cheap editions on the one hand or for the more expensive, attractively bound gift books on the other' (Lathey, 2012: 161).

Recent changes in children's (and their parents' and educators') reading habits, under the influence of technological advances may also be counted among the reasons why texts meant for the very young have been translated, retranslated, or adapted in various ways. Children, as the keenest users of electronic devices of all kinds and Internet almost-addicts these days, have become accustomed to resorting to the electronic medium both for getting informed and learning and for getting entertained. Their parents and educators, on the other hand, frequently find accessing online resources less time-consuming and more at hand when they read to them for any of these purposes. And, those who offer informative, learning, and entertainment online material targeted at children – often the result of a translation and adaptation process – upload what they think the users of such material need, in the form they consider most appropriate.

This chapter focusses on and compares cases of omissions and additions occurring in two Romanian variants of the English tale *Jack and the beanstalk*, as recorded by Edwin Sidney Hartland, and published in London in 1890 (of the numerous versions of the tale that have been known since its first recording by Joseph Jacobs in 1860, the Romanian readers seem to be most familiar with Hartland's). One of the two texts in Romanian is provided by a professional translator and university English teacher, Tamara Lăcătuşu, and was published in hardcover book format by Junimea, in 1981. The other is a version available online, posted by a person called Zina, whose professional identity is not disclosed.

What this analysis seeks to highlight is the effect that omissions and additions in the Romanian texts may have on their readers, thus moving one step forward from the mere identification of omission and addition as two of a range of translation options.

Under different labels and from different perspectives, omission and addition are dwelled on in all major taxonomies of translation methods, procedures or techniques (Nida, 1964, 1982; Newmark, 1988; Vinay & Darbelnet 1995; Venuti, 1998; Delisle *et al.*, 1999; Baker, 2002; Bassnett, 2002, etc.). Though to a much lesser extent, the reasons why they may be employed have also been considered by some other researchers (Dimitriu, 2012). However, what the actual consequences of their use may be at the level of the target text expressiveness and perlocutionary force has apparently remained of minor concern so far. Therefore, pointing at such consequences, even if in a small-scale analysis like this, brings in a missing link in the chain of 'what translation options are available to choose from – why and when they are used – what the consequences of using them may be on the target text itself, on the one hand, and on its readers, on the other'.

Within the framework of this analysis, omission is understood as not translating units of the source text and addition, as inserting in the target texts units that are absent from the source text. Omissions and additions that are triggered by differences in the grammatical systems of the two languages are not considered here.

9.2 Omissions and Additions in *Jack și vrejul de fasole* (Translation by Tamara Lăcătușu, Junimea 1981)

Omissions and additions in this Romanian variant of the tale occur at the word, phrase, and clause level. Their types and the potential consequences of their use are detailed as follows:

- omission of noun phrases and adverbs functioning as adverbials of manner:

	Source text	**Target text**
1)	... she kicked the beans away *in great passion.*	... ea le aruncă cât colo. ['... she kicked them *0* as far as she could.']
2)	Jack crept *softly* from his hiding place.	Jack se furișă din ascunzătoarea sa. ['Jack *crept 0* from his hiding place.']
3)	Jack... observed the giant counting over his treasures... he *carefully* put them in bags again.	Jack ... îl văzu pe uriaș numărându-și comorile, după care le vârî din nou în saci. ['Jack ... saw the giant counting his treasures, after which he *0* put them in bags again.']

4)	... the giant offered to restore her to liberty, on condition that she would *solemnly* swear that she would never divulge the story of her wrongs to anyone.	... uriașul se oferi să-i redea libertatea dacă jură că nu va spune nimănui despre toate relele pe care i le făcuse. ['... the giant offered to restore her to liberty if she *0* swears that she will not tell anybody about all the bad things that he had done to her.']

– omission of adverbs functioning as intensifiers:

	Source text	**Target text**
5)	The good woman at last suffered herself to be persuaded, for her disposition was *remarkably* compassionate.	Buna femeie se lăsă convinsă în cele din urmă, căci era miloasă din fire. [The good woman let herself be persuaded in the end, since she was *0* compassionate by nature.]

The adverbs that were ignored completely in the target text, as illustrated in the examples (1) to (4), are mainly carriers of the propositional meaning, i.e., they refer to something in the real world. However, they are not completely devoid of what Cruse (1986) calls 'expressive meaning', i.e., they indicate a certain attitude on the part of their user, which I identify as emotional involvement with the actions described. In (5), the intensifier adverb 'remarkably' has solely an expressive function, since 'removing it would not alter the information content of the message but would... tone its forcefulness down considerably' (Baker, 2002:14).

Thus, the absence of these adverbials in the target text leaves its core meaning intact, but diminishes its expressive potential and therefore, its capacity to emotionally engage the readers, to stimulate their imagination to work on long after the last word has been read. Kicking something 'in great passion' or swearing 'solemnly' is obviously much more intense emotionally than simply 'kicking' or 'swearing'.

– omission of entire clauses:

	Source text	**Target text**
6)	However, *the giant started up suddenly*, and not withstanding all his wife could say, he searched all round the room.	De data aceasta însă, uriașul, neluând în seamă vorbele femeii, scotoci prin toată casa. ['This time, however, the giant, *0* not taking into consideration the woman's words, searched all round the house.']

Clause omissions may also negatively influence the potential of the target text to stir its readers' sensitive involvement in the development of the tale, though for a different reason. In (6), no expressive meaning can be attached to the clause that is deleted in Romanian. However, not transferring it into the target text slows down the rhythm of the tale – both the verb 'to start up' and the adverb 'suddenly', indicating rapid movement and playing an obvious role in creating dynamism. This particular type of omission may itself have consequences on the degree of the readers' emotional participation in the tale. The fast progression of events, a genre peculiarity in the case of children's tales, keeps readers alert. Slowing down the rhythm goes hand in hand with decreasing the readers' alertness and emotional reactions.

The negative effect of such omissions is, sometimes, compensated for by:

- the addition of adjectives functioning as noun modifiers:

	Source text	Target text
7)	... *she* was obliged to support herself and him by selling everything she had.	... *biata femeie* se văzu obligată să câștige existența amândurora, vânzând tot ce avea. ['... *the poor woman* found herself forced to earn the living of both, by selling everything she had.']

- the addition of a syntactic construction, which is an equivalent of the superlative degree to an adjective used in the positive degree in the source text:

	Source text	Target text
8)	My punishment was a suspension of my power for a limited time, *an unfortunate circumstance...*	Drept pedeapsă, mi s-a luat puterea un timp, *o întâmplare* cum nu se poate mai *nefericită*.... ['As punishment, my power was taken away from me for some time, *an extremely unfortunate situation...*']

- the use of a binomial phrase, with an emphatic value, instead of a one word neutral equivalent of some English lexeme:

	Source text	Target text
9)	... reached the door *in safety* and soon arrived at the bottom of the beanstalk.	... ajunse la ușă și în curând se văzu *teafăr și nevătămat* lângă vrejul de fasole. ['... reached the door and soon found himself *safe and sound* next to the beanstalk.']

– the addition of entire clauses:

	Source text	Target text
(10)	He instantly set out and after climbing for some hours, reached the top of the beanstalk.	*În zadar îl imploră biata femeie să renunțe*, că el porni pe dată ... [‘*In vain did the poor woman beg him to give up*, he set out immediately...’]
(11)	Distracted at the absence of your parents, the servants went in search for them, but no tidings of either could be obtained.	Neliniștiți de absența părinților tăi, servitorii au pornit în căutarea lor. *Totul a fost însă în zadar*, pentru că nu au putut afla nimic despre stăpânul lor. [‘Distracted at the absence of your parents, the servants went in search for them. *But everything was in vain*, because they could not find anything about their master.’]

An expressive meaning component is evident in the case of all the added units in examples (7) to (11): the adjectives ‘biata’ (‘poor’) and ‘nefericită’ (‘unfortunate’), accompanied by the superlative marker ‘cum nu se poate mai’ (‘extremely’), the binomial ‘teafăr și nevătămat’ (‘safe and sound’), in which two synonyms are brought together and emphasis is, thus, obtained, as well as the two clauses ‘În zadar îl imploră biata femeie să renunțe’ (‘In vain did the poor woman beg him to give up’) and ‘Totul a fost însă în zadar’ (‘But everything was in vain’). They relate to the author’s (in our case, the translator’s) feelings and at the same time, heighten the readers’ emotional participation in the development of the tale. Thus, the level of text expressiveness is higher in the parts where the units quoted are added.

Arousing the emotions of a very young readership is, however, an aim that should be pursued with great care, and situations in which the text stirs children’s feelings and imagination in a direction that may harm them emotionally should be avoided. It is such avoidance that the Romanian translator seeks when resorting to:

– omission of single words related to taboo topics, in our case – death (of a parent):

	Source text	Target text
(12)	... you must persevere in avenging *the death* of your father	... trebuie să perseverezi în a ți răzbuna tatăl. [‘... you must persevere in avenging *0* your father.’]

– omission of entire clauses that deal explicitly with taboo topics (the same as aforementioned):

	Source text	Target text
(13)	... and the giant fell headlong into the garden. *The fall instantly killed him.*	... uriașul căzu lat în grădină ['... the giant fell flat in the garden *0*.']

A too strong and potentially damaging emotional effect on the child readers is minimized through these translation options. The information that is omitted in (12) and (13) can be recovered from the context so that the propositional content of these sections of the story remains unaffected.

– omission of noun phrases with negative connotations:

	Source text	Target text
(14)	... several poor victims whom the cruel giant reserved *in confinement* for his voracious appetite.	... victime pe care uriașul le ținea ca rezervă pentru nepotolita-i poftă de mâncare. ['... the victims that the giant kept *0* as a reserve for his voracious appetite.']
(15)	... or you will not prosper *in any* of your undertakings.	... altfel nu vei reuși în încercările tale. ['... otherwise you will not succeed in *0* your undertakings.']

In (14), the noun 'confinement' openly refers to cruel treatment applied to the giant's victims (who, as one finds out from the tale, have an unfortunate fate), while in (15), the use of the indefinite pronoun 'any' in the noun phrase 'in any of your undertakings' leaves the impression of no choice, of no way out of a curse. Omitting 'in confinement' and 'any' in the Romanian translation softens the original so that there may be no chills sent down children's spine when reading these parts of the text.

Besides the omissions already discussed, whose potential effects have been pointed out, one more deletion could be identified – that of a noun phrase functioning as an adverbial of time in the source text:

	Source text	Target text
(16)	There lived a poor widow, whose cottage stood in a country village a long distance from London, *for many years.*	Trăia odată o văduvă săracă a cărei colibă se afla într-un sat departe de Londra. ['There lived once a poor widow whose cottage was in a village far away from London *0*.]

The Romanian translator counterbalanced the omission of the adverbial of time 'for many years' – a translation choice with a negligible effect at the propositional level and no effect at all at the expressive level of the text – by the addition of the adverbial of time 'odată' ('once'):

	Source text	**Target text**
(17)	There *O* lived a poor widow, whose cottage stood in a country village a long distance from London, for many years.	Trăia *odată* o văduvă săracă a cărei colibă se afla într-un sat departe de Londra. ['There lived *once* a poor widow whose cottage was in a village far away from London.']

This addition that makes up for the omission turns the opening of the Romanian variant of the tale into a beginning that should be easily recognized by readers as genre-specific ('a fost *odată* (ca niciodată')-- 'once upon a time'). 'Odată' ('once') has, in Cruse's (1986) terms, an evoked, rather than a propositional meaning here. Once the beginning is identified as a characteristic of tales, the phatic function of 'Trăia odată...' ('There lived once...') is fulfilled – the channel of communication is opened and the readers are set in the proper mood for immersing into the story.

9.3 Omissions and Additions in *Jack și vrejul de fasole* (Online Version Posted by Zina)

The online version of the tale under scrutiny is drastically reduced as compared with the original. It is no longer a translation proper of Hartland's text, but rather a retelling in another language.

The reduced dimensions of the tale in the electronic format have been obtained by obvious omissions (no additions at all have been identified). However, once we admit that this is not a translation, but an adaptation of the source text, the omissions cannot be tracked in the same way as in the printed version analyzed in the previous section, i.e., at the corresponding word, phrase or sentence level. Instead, a global view of omissions may be a more appropriate analysis option here.

The shortening of the text is the result of two main types of reduction – through omission of whole fragments of various lengths and through reformulation of text parts of which certain details are deleted. Examples are as follows:

- omission of whole text fragments (no equivalent of the following fragments quoted can be identified in the target text):

(18) 'Now, mother', said Jack, 'I have brought you home that which will make you rich'.

(19) She... said, 'I will reveal to you a story your mother dare not. But before I begin, I require a solemn promise on your part to do what I command. I am a fairy and unless

you perform exactly what I direct you to do, you will deprive me of the power to assist you, and there is little doubt but that you will die in the attempt'.

(20) Jack set himself pensively upon a block of stone and thought of his mother. His hunger attacked him, and now he appeared sorrowful for his disobedience in climbing the beanstalk against her will, and concluded that he must now die for want of food.

(21) Looking upward, he could not descry the top. It seemed to be lost in the clouds. He tried it, discovered it firm and not to be shaken. A new idea immediately struck him. He would climb the beanstalk and see to whence it would lead. Full of his plan, which made him forget even his hunger, Jack hastened to communicate his intention to his mother.

(22) First, they passed an elegant hall, finely furnished. They then proceeded through several spacious rooms, all in the same style of grandeur, but they looked to be quite forsaken and desolate. A long gallery came next. It was very dark, just large enough to show that instead of a wall on each side, there was a grating of iron, which parted off a dismal dungeon, from whence issued the groans of several poor victims whom the cruel giant reserved in confinement for his voracious appetite. Poor Jack was in a dreadful fright at witnessing such a horrible scene, which caused him to fear that he would never see his mother, but be captured lastly for the giant's meat.

– reformulation of text parts from which certain details are deleted:

	Source text	**Target text**
(23)	When Jack hastened home with the beans and told his mother, and showed them to her, she kicked the beans away in a great passion. They flew in all directions, and were extended as far as the garden.	Acasă, mama se înfurie când le văzu și îi dădu peste mână lui Jack. Boabele căzură pe pământ. ['At home, mother got very angry and slapped Jack on the hand. The beans fell on the ground.']
(24)	Early in the morning, Jack arouse from his bed, and seeing something strange from the window, he hastened downstairs into the garden, where he soon found that some of the beans had grown in root and sprung up wonderfully.	A doua zi de dimineață, Jack descoperi că boabele încolțiseră. ['Next morning, Jack dicovered that the beans sprouted.']

(25)	Looking round, he was surprised to find himself in a strange country. It looked to be quite a barren desert. Not a tree, shrub, house, or living creature was to be seen.	Acolo descoperi o pustietate cum nu mai văzuse vreodată. ['There, he discovered a deserted land of a kind he had never seen before.']
(26)	Your father was a rich man, with a disposition greatly benevolent. It was his practice never to refuse relief to the deserving in his neighborhood, but, on the contrary, to seek out the helpless and distressed.	Tatăl lui Jack fusese un om bogat și bun la suflet. ['Jack's father used to be a rich and kind man.']

Both these types of omissions bear consequences on the propositional, and at the same time, on the expressive meaning of the text. The information content is obviously reduced severely, without, however, reaching the point when it can prevent the readers from understanding what the major stages in the plot development are. The expressive meaning is negatively affected to a larger degree, in that, by omitting emotionally loaded words, phrases and entire sentences and by breaking the smooth, and yet, dynamic linking up of events, the text loses the potential for a kind of creativity that Hollindale (1997), quoted by Lathey (2006), calls 'the childness' specific of children's literature, i.e., 'the quality of being a child – dynamic, imaginative, experimental, interactive, and unstable'. Thus, the overall expressive potential of the text, and consequently, its potential to keep the readers interested and to entertain them successfully decreases considerably.

The evoked meaning recovered in the printed translation of the tale by the addition of the adverb 'odată', which I discussed at the end of section 2, is lost in the electronic version, too. The introductory sentence here reads 'Într-un sat, departe de Londra, trăia o văduvă cu fiul ei, Jack' ['In a village, far from London, there lived a widow with her son, Jack'], which gives no formal indication that the text we are going to read belongs to the tale genre. As I suggested previously, failing to convey evoked meaning results into a lower potential of the text to set its readers into the appropriate emotional mood for receiving the tale. At the emotional level, readers miss the start, so to say, and unfortunately, as the online text progresses, their chances to get compensation for this are very little.

Omitting fragments that contain dialogues or direct address (such as those in (18) and (19)) and turning others into indirect speech also contribute to cancelling genre-specific features (marked orality in this case), with unfortunate consequences.

9.4 Conclusion

The analysis of additions and omissions in the two Romanian translations of *Jack and the beanstalk* generates the following final remarks.

In the case of the printed translation, the omissions at word, phrase and sentence level have minor effects on the overall propositional meaning of the text, but they do influence its expressive meaning. On one hand, they contribute to cancelling part of this meaning, and thus, part of the emotional effect that the text may have on its readers (this absence of what Nida (1964) calls 'dynamic equivalence' is, however, felt rather locally, since it is attached to very small portions of the text). On the other hand, omissions sometimes function as a 'protective' translation technique, in that, the potentially harmful consequences that the use of certain taboo words may have on the child readers are eliminated. Additions counterbalance omissions and manage to compensate for the loss of expressiveness and lessened potential to dynamize the readers' emotions. At the macro level, the source and the target texts can be regarded as equivalent from a propositional, expressive, and evoked meaning perspective, which further means that the Romanian text may be considered a fully functional translation.

This is no longer the case of the Romanian online retelling of the tale. Large-scale omissions of whole text fragments and smaller-scale ones, of details in certain parts of the text, diminish the quantity of information provided, and thus, reduce the propositional meaning (though, as I have pointed out, the key content elements are preserved). The outcome of such omissions is much more serious at the level of expressive meaning, since a dramatic drop off in the number of emotionally loaded text units and at the same time, the cancellation of a captivating flow of the text influence the readers emotional reactions negatively (moreover, no compensation strategies are applied for the loss of emotional force). As Burns (1962: 70) states, what the readers are faced with is just 'a collection of words..., something static not something with a life of its own – a mosaic rather than a painting'.

Zohar Shavit (2006: 26) points out that 'the translator of children's literature can permit himself great liberties regarding the text... That is, the translator is permitted to manipulate the text in various ways by changing, enlarging or abridging it, or by deleting or adding to it', as long as two principles are adhered to: adjustments should turn the text into one that is considered useful from an educational point of view and they should operate on plot, characterization and language so as to meet 'the society's perceptions of the child's ability to read and comprehend'. However, considering that children's literature fulfills its formative function through entertainment, it may well happen that, when tales are manipulated too drastically (even if with good, educational intentions) and are retold in a way that strips them of their expressive force, they fail not only from an emotional point of view, but they also fail in their formative role. This should be born in mind by those offering retellings of children's tales.

Since the circulation of children's literature and of texts derived from it often involves the work of translators, they, in their turn, should be aware of the consequences of their translation choices on the emotional effect the texts they produce may have on their recipients.

This brief analysis may raise the awareness of both those retelling stories (whether in the language in which they were originally written or in a different one) and of translators in the directions indicated, and thus, gain pedagogical implications.

Further investigation may envisage the extent to which tales are made available in the virtual space in formats similar to the one discussed here, as compared to the extent to which they are available in print. The situation I anticipate is that they are much more numerous online than between printed book covers. If this point of view is proved true, then it may also be true that the electronic medium favours the production of such deviations from the traditional tale genre pattern, and as a result, its degradation.

References

Baker, M. (2002). In other words. A coursebook on translation, 9th edition. London: Routledge.
Bassnett, S. (2002). Translation studies, 3rd edition. London and New York: Routledge.
Burns, M. (1962). The work of the translator. In L. Persson (Ed.), Translations of children's books (pp. 68-94). Lund: Biblioteksjanst.
Cruse, D. A. (1986). Lexical semantics. Cambridge: Cambridge University Press.
Dimitriu, R. (2012). Omission in translation. In Perspectives. Studies in translatology (pp. 163-175). Oxford: Taylor and Francis Group.
Delisle, J., Lee-Jahnke, H., & Cormier, M. (Eds.). (1999). Translation terminology. Amsterdam: John Benjamins.
Hollindale, P. (1997). Signs of childness in children's books. Stroud: Thimble Press.
Lathey, G. (2006). The translation of children's literature. A reader. Clevedon: Multilingual Matters.
Lathey, G. (2012). The role of translators in children's literature. Invisible storytellers. London: Routledge.
Newmark, P. (1988). A textbook of translation. Hertfordshire: Prentice Hall International.
Nida, E. (1964). Toward a science of translating. Leiden: E.J. Brill.
Nida, E. (1982). The theory and practice of translation. Leiden: E.J. Brill.
Shavit, Z. (2006). Translation of children's literature. In G. Lathey (Ed.), The translation of children's literature. A Reader (pp. 15-25). Clevendon: Multilingual Matters.
Venuti, L. (1998). Strategies of translation. In M. Baker (Ed.), Routledge encyclopedia of translation studies (pp. 240-244). London and New York: Routledge.
Vinay, J.P., & Darbelnet, J. (1995). Comparative stylistics of French and English. A methodology for translation. Amsterdam: John Benjamins.

Primary sources

Printed version of the tale in Romanian: *** (1981). Jack şi vrejul de fasole. In Basme populare engleze. Iaşi: Junimea.
Online version of the tale in Romanian: http://www.lecturirecenzate.ro/2014/08/jack-si-vrejul-de-fasole/ (accessed June 20, 2015).

Daniel Dejica

10 A Standards-Based Contrastive Analysis of Online and Printed Technical Translations in Romanian

10.1 Introduction

A recent study on the status of the translation profession in the European Union (Pym, Grin, Sfreddo, Chan, 2013) mentions that in Romania, 'the Ministry of Justice lists 32,856 "certified translators and interpreters" (since anyone with a degree in languages can qualify)' (2013: 21). According to the same study, as compared with other EU countries, in Romania, the lists of authorised translators exceeds by far the potential market demands, the number of translators and interpreters representing about 25 times the potential demand for professionals (2013: 21-22).

Concerning the types of translations requested in the Romanian market, I could not find any exact figures. However, given the size and development of the industry and international trade, the emergence of multinational engineering companies or the advances in science and technology in the past two decades in Romania and in the EU in general, one could estimate that the majority of the translations requested in the market are technical translations. On a global scale, based on Kingscott's estimations (2002: 247), an overwhelming 90% of the translations are technical translations. Taking into account a small margin of error, one would rightfully say that the same percentage may be relevant for the Romanian translation market as well.

The relationship between the large number of certified translators in Romania and the estimated requirements for technical translations in the Romanian market should be apt. The current study presents the results of a standards-based analysis performed on a corpus of online and printed technical translations. In light of the findings, several comments and recommendations are made.

10.2 Standards. An Overview

In a recent survey (Dejica, 2015) at the PCTS9 Conference (Professional Communication and Translation Studies) in Timișoara, most of the translators present at the session declared that they were not aware of the existence or of the provisions of the international translation standards and that, they approached their translations tasks using the knowledge they accumulated during their studies in tertiary education or at their workplace. Indeed, even if most BA courses on translation theory and practice include topics on the importance and necessity of translation assessment, a few academic courses mention and even fewer detail translation standards. References to such standards are, however, available on online educational platforms (Logos - http://courses.logos.it/), on the official pages of professional organizations (Romanian

Translators Association – www.atr.org.ro), on the official pages of the organizations that created such standards (detailed below), or in the specialised literature (House, 1997; Schäffner, 1998), to mention just a few.

10.2.1 Definition and classification of standards

The International Organization for Standardization (ISO) defines a standard as 'a document that provides requirements, specifications, guidelines or characteristics that can be used consistently to ensure that materials, products, processes and services are fit for their purpose' (www.iso.org).

Organizations like CEN, the European Committee for Standardization (https://www.cen.eu/), CENELEC, the European Committee for Electrotechnical Standardisation (http://www.cencenelec.eu/), ETSI, the European Telecommunications Standards Institute (http://www.etsi.org/), or Standards Australia (http://www.standards.org.au/) also provide similar definitions: they all share the idea that a standard is a document or a set of documents which provides rules or guidelines for performing some activities and achieving certain results, and which are approved and recognized by certain bodies. Concerning their classification, standards may be international, European, national or regional; each such class being adopted by international, European, national or regional bodies.

10.2.2 Translation Standards

Some of the most relevant translation standards include ISO 2384 'Documentation – Presentation of translations'; ISO 9001: 2008, 2015 'Quality management systems – Requirements'; EN 15038 'Translation Services – Service Requirements'; DIN 2345 'Translation Contracts'; UNI 10574 'Definition of services and activities of translation and interpreting enterprises'; ÖNORM D 1200 'Translation and interpretation services. Translation services. Requirements for the service and the provision of the service'.

10.2.2.1 ISO 2384 'Documentation – Presentation of translations'

The official description of the standard, as it appears in the online presentation abstract, reads that the standard 'sets out rules to ensure that translations are presented in a standard form which will simplify their use by different categories of user. Applies to the translation of all documents, whether the translation is complete, partial or abridged. Four types of translation are discussed.' (http://www.iso.org/).

Released in 1977, the standard was adopted unanimously by twenty-four countries, including Romania. The four types of translations discussed in the standard include books, periodicals, articles and patents. Regarding the presentation of translations, the standard details and clarifies aspects related to the structure of the original (content, paragraphing, subdivisions into clauses); notes and bibliographical references; formulae, equations, symbols, units; figures, legends, titles of figures and tables; transliteration; names and symbols of organizations; abbreviations; terminology; identification of authors; retranslation; geographical names; dates; translation of periodicals; name of the translator; and authority to publish translations.

10.2.2.2 ISO 9001: 2008, 2015 'Quality management systems – Requirements'

A general description of the ISO 9001: 2008 standard, updated in 2015, reads that it is 'designed to help organizations ensure they meet the needs of customers and other stakeholders while meeting statutory and regulatory requirements related to a product' (https://en.wikipedia.org/). Even if this is not a translation standard, it is relevant for the language industry in that it certifies that the language service provider follows a process (http://www.languagescientific.com/). The requirements of the ISO 9001: 2008, 2015 certification are based on a set of principles, which include customer focus, leadership, involvement of people, process approach, system approach to management, continual improvement, factual approach to decision making, and mutually beneficial supplier relationship.

10.2.2.3 EN 15038 'Translation Services – Service Requirements'

The scope of this standard is to specify 'the requirements for the translation service provider (TSP) with regard to human and technical resources, quality and project management, the contractual framework, and service procedures' (EN 15038, 2006: 6). EN 15038 is a European Standard which has the status of a national standard. Twenty-nine countries, including Romania, have adopted it. (In Romania, it is identified under SR EN 15038.) As stated in the Foreword section, 'any conflicting national standards shall be withdrawn at the latest by November 2006' (EN 15038, 2006: 3). This requirement applied to all twenty-nine countries, which were bound to implement it.

According to the International Federation of Translators (FIT), 'The purpose of the European Standard is to ensure high-quality translation services, fair competition, improved transparency and end user satisfaction.' (http://www.fit-europe.org/). FIT Europe, the Regional Centre of the International Federation of Translators, developed a set of Recommendations on Criteria for Conformity Assessment and Certification under EN 15038 (http://www.fit-europe.org/). The recommendations follow the numbering of the standard and consist of three main sections, including basic requirements (human resources, technical resources, quality management system, and project management), client–TSP relationship (Translation Service Provider), relationship (feasibility of the project, quotation, client–TSP agreement, handling of project-related client information, and project conclusion), and procedures in translation services (managing translation projects, preparation – including the administrative, technical and linguistic aspects – and the translation process).

10.2.2.4 Other Translation Standards

As mentioned above, EN 15038 cancelled any conflicting standards in the EU after November 2006. As of today, fifteen EU countries adopted EN 15039 as national standards. As listed in Wikipedia, these include: PKN EN 15038 Poland, BS EN 15038 Great Britain, DIN EN 15038 Germany, NF X50-670 France, ÖNORM EN 15038 Austria, SN EN 15038 Switzerland, UNE-EN 15038 Spain, SFS EN 15038 Finland, UNI EN 15038 Italy, EVS-EN 15038 Estonia, MSZ EN 15038 Hungary, SS-EN 15038:2006 Sweden, SR-EN 15038 Romania, BDS EN 15038 Bulgaria and LST EN 15038 Lithuania (https://en.wikipedia.org/).

Several European countries developed their own standards before 2006, including, for instance, Germany (DIN 2345), Italy (UNI 10574), or Austria (Önorm D 1200). A list of the translation standards between 1996 and 2006 is available in Stejskal (2009: 296).

The German Standard DIN 2345 was published in 1998 by the German Institute for Standardization. As shown in Samuelsson–Brown (2006), the standard 'contains different conditions for concluding contracts between translators and clients. Among the issues the standard considers are: the original (or source) text, choosing the translator, setting up a contract between client and translator, target text, proofreading' (2006: 137).

The Italian Standard UNI 10574 'Definition of services and activities of translation and interpreting enterprises' was published in 1996 and 'defines the requirements and procedures that providers of translation and interpretation services should implement in their daily activities' (http://www.translinknet.be/).

The Austrian standard ÖNORM D 1200:2000 'Translation and interpretation services. Translation services. Requirements for the service and the provision of the service' was published in 2000. It includes service requirements and provisions of the service. It details the types of competence required of the translators, including translating, linguistic and textual, research, cultural, and technical competence.

10.3 Technical Translation Research Today

The specialised literature dealing with LSP translation, in general, and technical translation, in particular, includes, but is not limited to Byrne (2006, 2012), Gerzymisch–Arbogast, Budin, Hann (1992), Hofer (2008), Hansen, Malmkiaer, Gile (2004), Kučiš (2013), Olohan (2016), Pârlog (2014), Sandrini (2006), Somers (1996), Vîlceanu (2009) or Wright (1993). Still, as Byrne (2006: 1) claims, 'this vocational and industrial type of translation has been largely neglected in the literature of translation theory'. Byrne supports his claim on a study conducted by Aixelá (2004), according to which, 'out of 20,495 publications listed in the BITRA multilingual bibliography of translation research only 1,905 or 9.3% addressed technical translation'.

10.3.1 Translation standards and technical translation

Up to my knowledge, studies on the relationship between translation standards and technical translation (compliance or non-compliance with standards, standards-based contrastive analyses between source texts and target texts or different languages, case studies, etc.) have not been published, which supports Aixelá's (2004) findings and proves that research in this field is still not fully explored and justifies the present research.

An analysis of the standards presented in section 2 reveals that they are mostly descriptive, rather than prescriptive: the focus is on *what* aspects should be followed, rather than on *how* to follow them in the translation process. In addition, the analysis reveals that most standards include organizational details related to human resources, client–translator relationship, forms of delivery, confidentiality, etc., and only few standards include recommendations or requirements related to the content or form of the translation.

In order to create a standards-based grid for the analysis of technical translations, I used a three-stage process, consisting of the following stages: 1) standards-based identification of relevant formal and content-specific aspects for translation, 2) overall classification of aspects, and 3) genre-based selection of relevant aspects.

The results of the standards-based identification of relevant formal and content-specific aspects for translation are presented in Table 10.1. Not all of the standards detailed earlier made clear reference to the form or content of translations; therefore, I listed only the requirements related to the form and content (mostly linguistic) of the translations present in the standards ISO 2384, EN 15038, and DIN 2345.

The overall analysis of the requirements listed in Table 10.1 allows for their classification in formal and content-specific aspects. The content-specific aspects may, in turn, be classified in language- and translation-specific requirements (Table 10.2).

The requirements in Table 10.2 are not exhaustive. When used as a basis for the assessment of translations, other aspects may be included, for instance, requirements or guidelines pertaining to the translation of cultural items or culturèmes. Since the current analysis is based only on the requirements existing in the standards, such additions will not be made here.

Table 10.1: Standards-based formal and content-specific translation requirements

Requirements	Standard		
	ISO 2384	**EN 15038**	**DIN 2345**
Structure of the original (paragraphs, numbering, subdivision of the text into clauses)	X		X
Notes and bibliographical references	X		
Formulae, equations, symbols, units	X		X
Figures, legends, titles of figures and tables	X		X
Transliteration	X		
Names and symbols of organizations	X		
Abbreviations	X		
Terminology	X	X	
Identification of authors	X		
Retranslation	X		
Geographical names	X		
Dates	X		
Translation of periodicals	X		
Name of the translator	X		
Authority to publish translations	X		
Grammar: syntax, spelling, punctuation		X	
Lexis: lexical cohesion and phraseology		X	
Style: compliance with the proprietary or client style guide, including register and language variants		X	X
Locale: local conventions and regional standards		X	
Formatting		X	
Target group and purpose of the translation		X	
Meaning has been conveyed		X	
Omissions		X	
Errors		X	
Footnotes			X
Appendices			X

Table 10.2: Overall classification of formal and content-specific translation requirements

Formal aspects	Structure of the original (paragraphs, numbering, subdivision of the text into clauses)
	Notes and bibliographical references
	Formulae, equations, symbols, units
	Figures, legends, titles of figures and tables
	Footnotes, endnotes
	Names and symbols of organizations
	Abbreviations
	Transliteration
	Identification of authors
	Retranslation
	Geographical names
	Dates
	Translation of periodicals
	Appendices
	Name of the translator
	Authority to publish translations
	Formatting
Content-specific aspects	**Language-specific**
	Grammar: syntax, spelling, punctuation
	Lexis: lexical cohesion and phraseology
	Terminology
	Style: compliance with the proprietary or client style guide, including register and language variants
	Locale: local conventions and regional standards
	Translation-specific
	Meaning (conveyed)
	Target group and purpose of the translation
	Omissions
	Errors

In a similar way, since standards refer to translation in general, not to the translation of a specific genre, certain aspects may be omitted from the list to suit particular genres' specificities (Bhatia, 1993; Swales, 1993; Trosborg, 2000; Dejica 2011). For example, in the case of technical translation, aspects which should be considered redundant or irrelevant would include geographical names or the translation of periodicals.

Table 10.3 contains a tentative standards-based grid, which I believe would be relevant for the analysis of the formal and content-specific aspects of technical translation. The content-specific aspects may also be classified into language-specific and translation-specific aspects:

Table 10.3: Tentative standards-based analysis grid for technical translations

Formal aspects	Structure of the original (paragraphs, numbering, subdivision of the text into clauses)
	Notes and bibliographical references
	Formulae, equations, symbols, units
	Figures, legends, titles of figures and tables
	Footnotes, endnotes
	Names and symbols of organizations (companies, manufacturers)
	Abbreviations
	Appendices
	Formatting
Content-specific aspects	**Language-specific**
	Grammar: syntax, spelling, punctuation
	Lexis: lexical cohesion and phraseology
	Terminology
	Style: compliance with the proprietary or client style guide, including register and language variants
	Locale: local conventions and regional standards
	Translation-specific
	Meaning (conveyed)
	Target group and purpose of the translation
	Omissions
	Errors

10.4 A Standards-based Analysis of Technical Translation

10.4.1 Corpus description

A classification of technical genres and sub-genres would include, but would not be limited to, instruction manuals, technical guides or user's manuals; technical brochures, technical product presentations; technical drawings; technical product catalogues; technical reports, technical memorandums; feasibility studies; technical projects; spare parts list, catalogues, technical forms; or technical correspondence.

A standards-based analysis of technical translation from a genre perspective, covering all or most of the genres or sub-genres mentioned above would be very useful, but would exceed the size of the present research. Accordingly, this research is narrowed to user's manuals, instruction manuals, and product descriptions, since I assume that the demand in the market for the translation of such genres is greater than the one for other technical genres mentioned above.

Another filter I applied in the process of narrowing and selecting the corpus for analysis refers to the range of products. I opted for various smart devices, digitally enabled, including phones, phablets, tablets, TVs, game consoles, watches, washing machines, etc., for the same reason: due to the market demands, such devices and their improved versions are released in the market more frequently and at shorter time intervals than other technical devices (a lawn mower, for example), and hence, the need for the translation of the user's manuals, instruction manuals and product descriptions, which accompany these products is greater.

Last but not least, in selecting the corpus, I opted for materials for products representing different brands, from different manufacturers, including LG, Philips, Samsung, Sony, or Treo, published at different time intervals in the last 10 years. Thus, the probability of analysing translations performed by different translators would be higher and the results more relevant.

The corpus included 45 interdependent translations (Sager, 1993, 177-182) in printed format and electronic format available online; the analysis and interpretation of the results are presented in the next section.

10.4.2 Analysis and interpretation of the results

The contrastive analysis was performed on the corpus described earlier and followed the observance or non-observance of the formal and content-specific aspects presented in Table 10.3.

When analysed individually, on the whole, the target texts look and read well; however, the contrastive analysis between the STs and the TTs shows that in most cases, the formal and content-specific requirements of the standards are not observed in most of the target texts.

10.4.2.1 Formal requirements

The structure of the original (paragraphs, numbering, and subdivision of the text into clauses) was respected in most of the cases.

Notes and bibliographical references were usually few and were translated in full.

Formulae and equations were not present in the source texts I analysed. The symbols like ©, ®, or ™ were preserved in the target texts in all of the cases. However, units seem to be approached in different ways by translators, most likely because of two reasons: as a marketing strategy (client's requirements), but also due to the permissive formulation of ISO 2384, according to which 'units are preferably not translated. If the units are converted, the value of the original units should be given in brackets'. In the case of smartphones, units were never translated or converted in Romanian (e.g., ST: *5 inch display* – TT: ecran de *5 inch* / ecran de *5 inci*). This is also the case of computer monitors (e.g., LG LED Monitor 22M37A, *21.5'* Full HD Black – TT: Monitor LED LG 22M37A, *21.5'*, Full HD, negru). However, the translation strategies differ in the case of flat TV screens: for smaller TV screen sizes, usually up to 32 inch, the original units were preserved (probably due to the similar size and the influence of computer monitors) and in some cases translated (TT: Televizor Smart LED LG, *80 cm*, 32LF580V, Full HD), but for wider TV screens, usually above 40 inch diagonally, units were never preserved, and they were always converted (e.g., TT: Televizor

OLED Curbat Ultra HD 3D, Smart TV, *139 cm*, LG 55EG960V). Indeed, when reading the product description online, a Romanian prospective buyer would probably find it easier to visualize a 139-cm TV screen than a 55 inch one, even if the two are identical in size. Another difference is related to the conversion of units. For the same size and unit (i.e,. a 32-inch TV screen), I could find three different screen sizes in the Romanian translations: 80, 81, and 82 cm, which is also most likely due to the way a certain brand is marketed. (The exact size in centimeters as given by asknumbers.com is 81,28 cm.)

Figures, legends, titles of figures and tables were translated in full. The text which accompanied them was also translated in full in all of the cases. However, the standard recommendation (ISO 2384) regarding the position in which they appear in the source texts was not followed in 35 of the 45 target texts analyzed. An example is shown in Figure 10.1, which presents in parallel the source text (English) and the target text (Romanian); the example is an excerpt from a user's manual, which details tips for washing and energy saving. As it can be seen, the form of the ST differs: 1) in the ST, the section starts at the top of the page, which is a common editorial practice for new sections or chapters; this detail is not followed in the TT; 2) the position of the table presenting the hardness scale and the degree of hardness differs and 3) the image representing the dispenser compartment is not next to the table, but on a different page, which would probably irritate the user since there is clearly a close relationship between the quantity of detergent used, presented in the table, and the dispenser compartment, which indicates where the detergent should be placed. Thus, the intention of the source text writer – to facilitate understanding by associating figures with images – is not preserved in the target text.

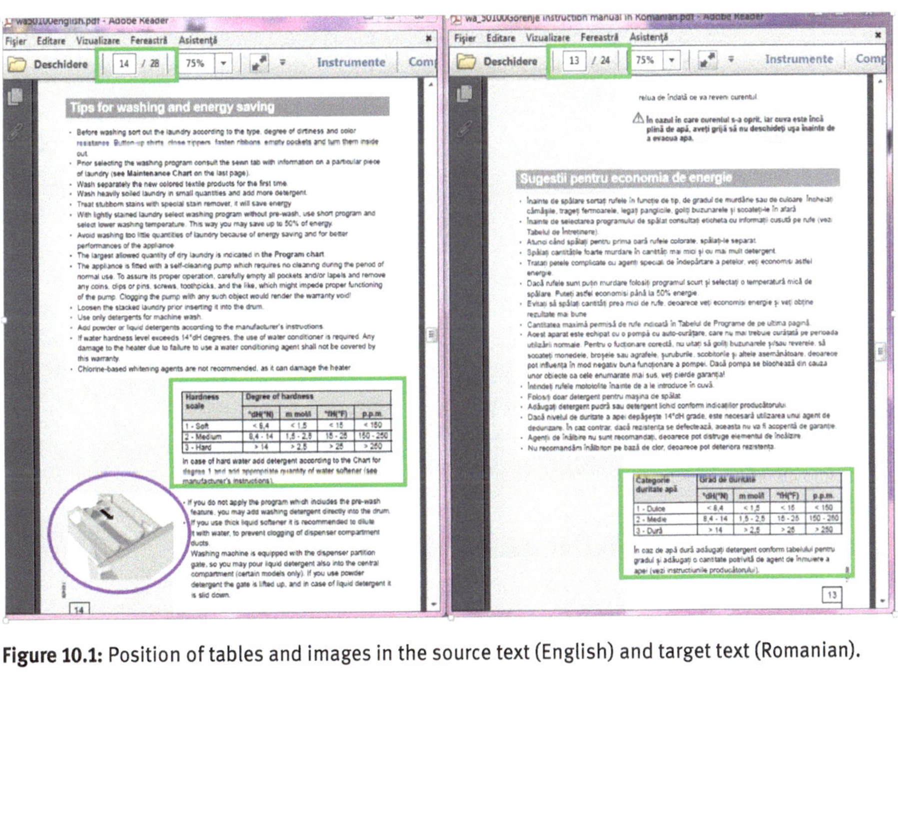

Figure 10.1: Position of tables and images in the source text (English) and target text (Romanian).

Almost the same differences between the position in the text can be observed in Figure 10.2, which represents an excerpt from the user's manual of a smartphone: 1) the TT does not start with the title of the new section at the top of the page and 2) the TT fails to preserve the same information at the bottom of the page. A contextual analysis of the ST would reveal that the intention of the ST writer was to contrast two different models of the same smartphone, and in doing so, s/he presented the position of the Sleep/Wake button in two adjacent pictures. The pictures and the accompanying text, all on the same page, were clearly meant to highlight the differences – which, in this way are obvious – and to facilitate understanding of the text. Since the TT does not preserve the same position of the images and the text – the second image and the accompanying text are on the following page, the ST would most likely fail to convey the intention of the TT writer. As I mentioned before, such examples are common for most of the target texts I analyzed. On the whole, failure to comply with the standard's recommendation on the position in the text would make it difficult to read the texts interdependently, would disrupt a carefully arranged document structure, and eventually would lead to a difference in focus in the target text.

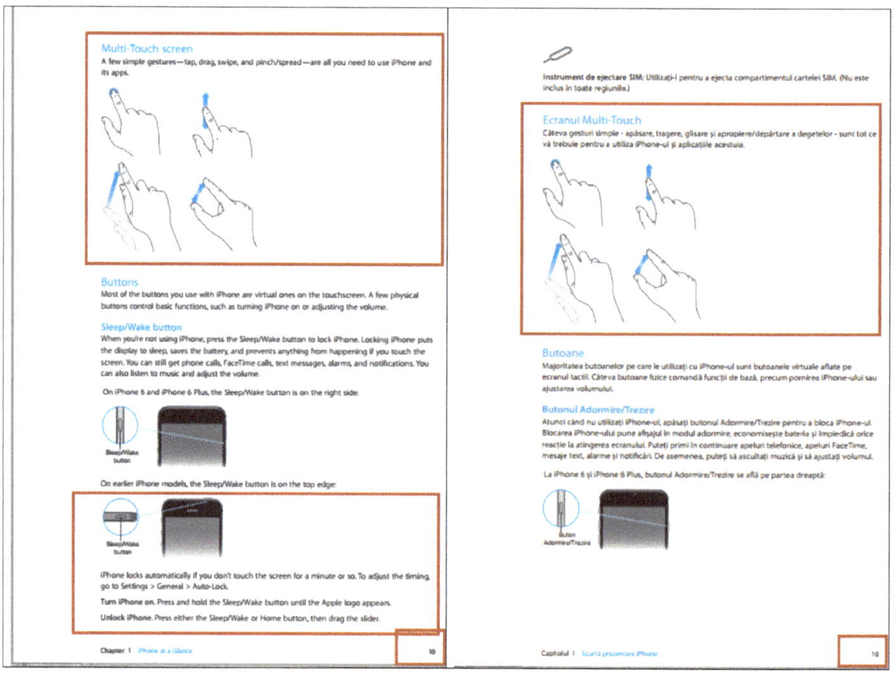

Figure 10.2: The position of images and of surrounding text in the source text (English) and target text (Romanian).

Footnotes and endnotes were not present in the technical texts I analysed.

The names and symbols of organizations (companies and manufacturers) were usually translated in full. However, in 10 target texts, there were cases where manufacturers' symbols or logos present in the source texts were missing. Some of the missing logos include the SD card symbol , the Mac OS symbol , or the Microsoft Windows OS symbol .

Abbreviations were translated using proper equivalents in Romanian, and appendixes were translated in full.

Formatting was not respected in 35 target texts. The cases of non-observance included different text alignment and font style. Page margins and font-size and types were preserved in the target texts.

10.4.2.2 Content-specific requirements

As suggested in Table 10.3, the content-specific aspects include language- and translation-specific aspects. The 45 texts which constituted the corpus totalled about 900 pages. A standards-based quantitative analysis would have exceeded the purpose of this research; hence, I performed a qualitative analysis on one page, randomly selected from each document, the amount of analysed text totalling 45 pages.

The *language-specific* grammatical analysis showed that there were no syntax errors in the target texts.

Concerning the spelling, in 6 out of the 45 excerpts analysed, the translator did not use any diacritical marks in Romanian, in contradiction with standards' requirements according to which translators should pay special attention to the use of diacritical marks in the target texts. Proper punctuation was not used in most of the analysed cases. Most common punctuation errors included misplaced comma use or faulty spacing before or after punctuation marks.

Lexical cohesion was preserved in all of the analysed excerpts.

In the case of terminology, the translator did preserve the specialised terms in the target texts; however, there were cases in which it would have been more appropriate to use a target language equivalent instead of a borrowed term. In few cases I could identify the use of barbarisms in Romanian, one such example being *Organizer Palm OS* instead of *Organizator Palm OS* for the English *Palm OS Organizer.* In other cases, terms were not translated at all (exemplification and discussion as follows).

There were no identifiable differences in terms of style or register.

The *translation-specific* analysis showed that meaning had been conveyed in all of the analysed texts. A particular case is represented by the translation of terminology and its relation to text understanding and meaning, which deserves special attention. As presented earlier, in most of the cases, terms were borrowed or translated using proper target language equivalents. However, I could identify two more different cases, which could be interpreted as different overall approaches to the translation of the source texts. In the first case, English terms were used in the Romanian texts as such, without further explanation or additional translation (Figure

10.3). In the second case, English terms were used in the Romanian texts, followed by the Romanian translation (Figure 10.4).

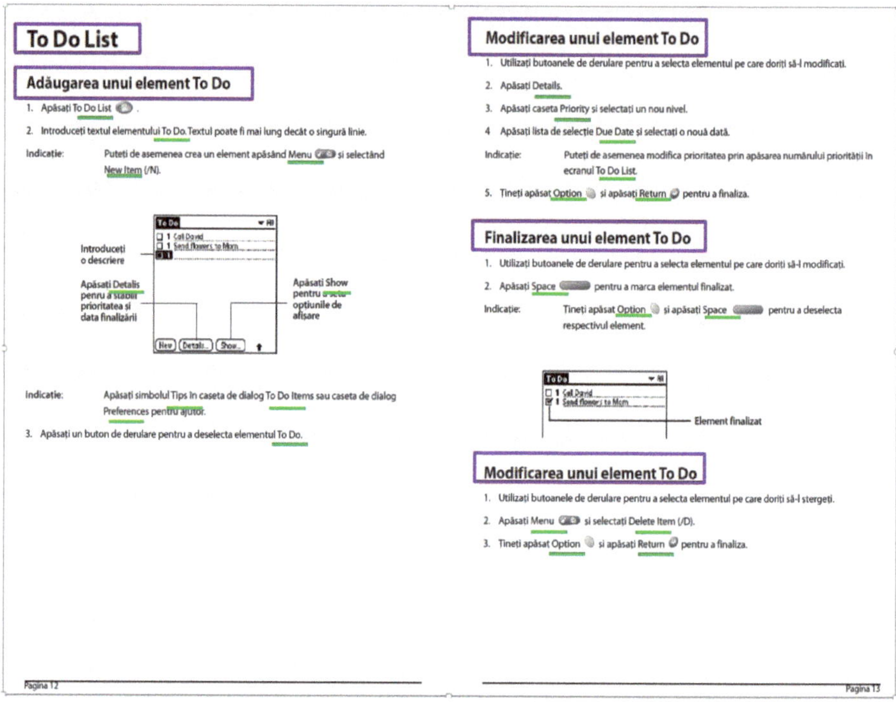

Figure 10.3: Example of a target text (Romanian) which preserves the source text terms (English).

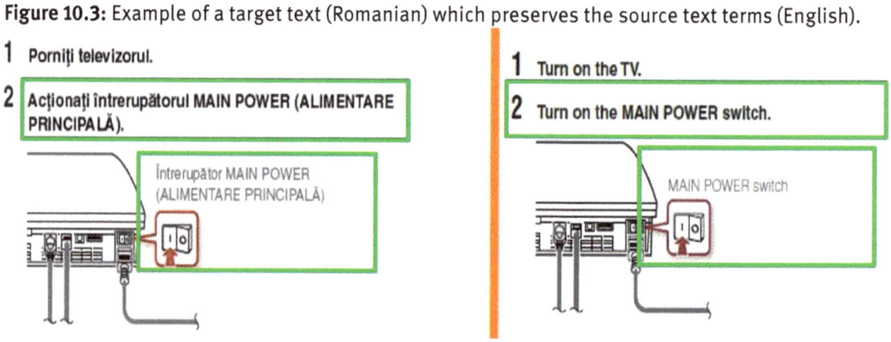

Figure 10.4: Example of a target text (Romanian) which preserves the source text terms (English) and provides their translation in brackets.

Figure 10.3 is taken from a user's manual and represents the translation of the operating system of a mobile phone. The translator's choice of preserving the English terms (23 occurrences on two pages, including terms like *to do list*, *delete*, *new item*, *option, priority, return, space*, etc.) in the Romanian text may be justified by the fact that the mobile phone did not include Romanian as an option for the selection of the language of installation of the operating system. Even if justifiable, the translator's decision to preserve the English terms in Romanian makes the text readable only by bilinguals or by English language speakers. In this case, even if the meaning is conveyed, text understanding is difficult, and for some people, even impossible. A glossary including the translation of terms or their translation in brackets would have been appropriate solutions. This latter solution is actually the case with the second example (Figure 10.4), which shows a text taken from a TV user's manual. The translator kept the English terms into Romanian, but offered their Romanian translation, facilitating, thus, text understanding.

In my analysis, I took for granted that the translation situation required preserving the same characteristics of the target group, purpose of the translation or intention of the writer in the target texts, as well, and I did not find any differences in this respect.

No significant omissions or errors which would have changed the meaning of the translation were identified.

10.5 Conclusion

This research offered the opportunity to reach several conclusions and to identify several research questions, which need further investigation.

An analysis of the standards presented in section 2 reveals that they are mostly descriptive, rather than prescriptive. Moreover, in many cases, the standards' requirements may be interpreted as recommendations (extensive use of *should*). If some standards are relatively new (EN 15038, 2006), others are rather old (ISO 2384, 1977), and as compared with the other standards, have never been updated. Given that ISO 2384 was published before the invention and implementation of the World Wide Web, and that nowadays most technical translations are available online or only in an electronic format, translation professionals would rightfully question themselves as to the relevance of the standard's requirements for today's translation situations. The answer would become even more obvious in the case of digital-born translations and would definitely require the immediate attention of bodies responsible with the maintenance of such standards.

There is no doubt that translation standards are necessary and useful. When analysed individually, most technical target texts read and look well. However, the standards-based formal and content-based contrastive analysis revealed that there are cases where the provisions of the standards are not followed. In such cases, the target texts' qualities and properties are not met (in particular, consistency and adequacy), which in turn, leads to an overall poor quality of the translations. The possible reasons include translators' unfamiliarity with the standards, lack of using dedicated software, or even the fact that such translations are performed by non-professionals. Given the market demand for technical translations and implicitly for well-trained translation professionals, this is a luxury no one should afford.

There should be a correlation between translation ethics, functional translation or functional nature of a standard, and observance of standards' requirements. Functionalism in a standards-based translation context should not be understood as the possibility to gamble with the text, to alter it unconscientiously or according to unethical requests (e.g., avoid transferring company or manufacturers' logos in the target text, misplacing the position of tables or figures, giving inadequate equivalents for technical terms, or non-observance of other standards' provisions), but as the possibility to tailor a translation according to the client's requirements, by following standards' or any other ethical requirements.

In the context of preparing translators for the translation of technical texts in the digital age, universities and professional organizations in Romania should be more visible and have an increased role. Some of the immediate measures may include developing and accrediting technical translator-training programmes, setting up life-long learning programmes (including, but not limited to masters' programs in technical or specialized translations, intensive courses, summer schools), and last but not least, raising awareness of the importance of using dedicated software to improve the quality and efficiency of technical translations. Other supportive measures may include setting up standardised examinations for technical translators' certification or creating periodical re-evaluation programmes for technical translators.

This study has also confirmed that the field of technical translation offers many opportunities for further research, particularly in the context of its development and evolution in the digital era. Some of the research questions and topics that emerged from this study and which deserve attention include, in a random order, 1) the evolution of online technical genres and their implications for translation, 2) the identification and analysis of the features of online interdependent technical translations, 3) the possibility of creating standardized genre-based requirements for the translation or assessment of technical translations (or of specialized translation in general), 4) the relationship between different translation schools and the nature of translation standards, or 5) the relevance of the existing translation standards for digital-born translations and the necessity or opportunity of updating standards to be one step ahead or at least keep up with the continuous evolution and form of translations in the digital age.

References

Aixelá, F. J. (2004). The Study of Technical and Scientific Translation: An Examination of its Historical Development. JoSTrans - The Journal of Specialised Translation: Volume 1, Number 1.

Bhatia, V.J. (1993). Analysing Genre. London: Longman.

Byrne, J. (2006). Technical Translation. Usability Strategies for Translating Technical Documentation. Springer: Dordrecht.

Byrne, J. (2014). Scientific and Technical Translation Explained: A Nuts and Bolts Guide for Beginners. Routledge: New York.

Dejica, D. (2011). Identifying and analysing professional genres' peculiarities for translation purposes: a methodological approach'. In Frentiu, L. (ed.), Romanian Journal of English Studies, 8/2011. Timisoara: Editura Universitatii de Vest, 155-166.

Dejica, D. (2015). ISO 2384: The Case of Technical Translations in Romanian. Presentation given at the International Conference PCTS9, Professional Communication and Translation Studies, March 26-27, Timisoara, Romania.

DIN 2345. (1998). Translation Contracts.

EN 15038. (2006). Translation services – Service Requirements.

Gerzymisch-Arbogast, H. & Budin, G., Hofer G. (eds.) (2008). LSP Translation Scenarios. MuTra Journal, published by the ATRC Group.

Hann, M. (1992). The Key to Technical Translation. John Benjamins: Amsterdam/Philadelphia.

Hansen, G. & Malmkiaer, K., GIle, D. (2004). Claims, changes and challenges in translation studies: selected contributions from the EST Congress, Copenhagen 2001. John Benjamins: Amsterdam / Philadelphia.

House, J. (1997). Translation Quality Assessment. A Model Revisited. Tübingen: Narr.

Kučiš, V. (ed.) (2013). Translation in Theorie und Praxis. Frankfurt am Main, Berlin, Bern, Bruxelles, New York, Oxford, Wien.

ISO 2384. (1977). Documentation – Presentation of translations.

ISO 9001. (2008). Quality management systems – Requirements.

ISO 9001. (2015). Quality management systems – Requirements.

Kingscott, G. (2002). Technical Translation and Related Disciplines. In: Perspectives: Studies in Translatology. Vol. 10:4, 247-255.

Olohan, M. (2015). Scientific and Technical Translation. Routledge: London and New York.

ÖNORM D 1200. (2000). Translation and interpretation services. Translation services. Requirements for the service and the provision of the service.

ÖNORM D 1201. (2000). Translation and interpretation services. Translation services. Translation contracts.

Pârlog, A.-C. (2014). A General Approach to Technical Translations. In Chirimbu, S. & Peter, K., Burda, A. (Eds.), Procedia of Interdisciplinarity. DSCEI 2014 International Conference, Selected Papers, Section: Humanities. Utah: Aardvard Global Publishing, 251-259.

Pym, A., & Grin, F., Sfreddo, C, Chan A. (2013). The Status of the European Profession in the European Union. London, New York, Delhi: Anthem Press.

Sager, J. C. (1993). Language Engineering and Translation. Consequences of automation. Amsterdam: Benjamins.

Samuelsson-Brown, G. (2006). Managing Translation Services. Cleveland, Buffalo, Toronto: Multilingual Matters LTD.

Sandrini, P. (2006). LSP Translation and Globalization. In Gotti, M. & Šarčević, S. (Eds.), Insights into Specialized Translation. Linguistic Insights 46. Bern Berlin Frankfurt: Peter Lang, 107-120.

Schäffner, C. (1998). Translation and Quality. Clevedon: MultilingualMatters.

Somers, H. (Ed.) (1996). Terminology, LSP and translation: studies in language engineering in honour of Juan C. Sager. John Benjamins: Amsterdam/Philadelphia.

Stejskal, J. (2009). Quality Assessment in Translation. In Forsner, M., & Lee-Jahnke, H. Schmitt, P.A. (Eds.), CIUTI-Forum 2008. Enhancing Translation Quality: Ways, Means, Methods. Bern: Peter Lang.

Swales, J. (1993). Genre Analysis. Cambridge: Cambridge University Press.

Trosborg, A. (2000). Text and Genre Analysis for Translators: Structure, Content and Function of the Model. In Schaffner, C. (ed.), Current Issues in Language and Society, Vol 7, No. 3. Multilingual Matters: Clevedon, England, 229-244.

UNI 10574. (1996). Definition of services and activities of translation and interpreting enterprises.
Vîlceanu, T. (2009). Problematica standardizării terminologice în ştiinţa şi tehnologia de vârf in Actele Colocviului Internaţional Teoria, practica şi didactica traducerii specializate, Craiova. Uninea Latină / Union latine 05/2009, 170-175.
Wright, Sue Ellen, Leland D. Wright, Jr. (Eds.). (1993). Scientific and Technical Translation. ATA Scholarly Monograph Series VI. John Benjamins: Amsterdam/Philadelphia.

Webography

CEN. The European Committee for Standardization. https://www.cen.eu/
CENELEC. The European Committee for Electrotechnical Standardisation. http://www.cencenelec.eu/
CENELEC. What is a European Standard (EN)? Available at http://www.cencenelec.eu/standards/DefEN/Pages/default.aspx. Accessed June 2015.
EN 15038. Available at https://en.wikipedia.org/wiki/EN_15038. Accessed June 2015.
ETSI. The European Telecommunications Standards Institute (http://www.etsi.org/)
ETSI. What are standards? Available at http://www.etsi.org/standards/what-are-standards . Accessed June 2015.
European Committee for Standardization. European Standards. Available at https://www.cen.eu/work/products/ENs/Pages/default.aspx. Accessed June 2015.
FIT Europe. Recommendations on Criteria for Conformity Assessment and Certification under EN 15038. Available at http://www.fit-europe.org/vault/RecAssessCriteria.pdf. Accessed June 2015.
FIT Europe. What is the European Standard EN 15038 and How does it help translators? Available at http://www.fit-europe.org/en/what-we-do/completed-projects/standard-en-15038#terminology. Accessed June 2015.
International Standardization Organization. ISO 2384:1977. Documentation – Presentation of Translations. http://www.iso.org/iso/catalogue_detail.htm?csnumber=7274 . Accessed June 2015.
International Standardization Organization. What is a standard? Available at http://www.iso.org/iso/home/standards.htm. Accessed June 2015.
ISO 9000. Available at https://en.wikipedia.org/wiki/ISO_9000. Accessed June 2015.
Logos. The International Standard ISO 2384. http://courses.logos.it/EN/1_1.html
McGowan, J. (2013). What Is ISO 9001 Certification, And How Does It Relate To Translation? Available at http://www.languagescientific.com/language-services-blog/iso9001-certification-and-how-it-relates-to-translation.html Accessed July 2015.
Romanian Translators Association. www.atr.org.ro
Standards Australia (http://www.standards.org.au/)
Standards Australia. What is a Standard? Available at http://www.standards.org.au/standardsdevelopment/what_is_a_standard/ . Accessed June 2015.
Translation Standards. Available at http://www.translinknet.be/quality/standards.html. Accessed June 2015.

Violeta Tănase

11 Extratextual Elements in Subtitling – The Battle of Linguistic and Cultural Codes

11.1 Introduction

The translation of the audiovisual text has been dealt with extensively within the discipline of Translation Studies in the past decades, to the point where it would not be an overstatement to say that audiovisual translation has become a branch in its own right of the discipline of Translation Studies, having its own specific terminology, concepts, approaches and gaining ever increasing academic representation. However, in spite of the huge amount of literature in the field and the increasing number of scholars and academics interested in audiovisual translation, few issues regarding audiovisual translation have been exhaustively discussed. This is partly because of the interdisciplinarity of the field, the multimodal nature of the text, the diversity of audiovisual translation modes, and also due to the fast technological changes that constantly open new possibilities in the field of audiovisual translation.

In order to understand the importance of extratextual markers in the translation of audiovisual texts, it is useful to revisit this particular type of text in terms of structure and diversity.

11.2 Structure of the Audiovisual Text

The source language text subject to translation in the audiovisual text (AVT) has a complex structure, being made up of several codes that function simultaneously to produce a single desired effect. Delia Chiaro (2009, 142) speaks about the translation of screen products (SP) intended primarily for being seen and made up of a complex visual code and an acoustic code. The visual code contains elements such as: the actor's movements, gestures and facial expressions, the scenery, the costumes, lightning and colour and written verbal information rendered through signposts, street signs, banners, newspapers, letters, notes, etc. The acoustic code has, in turn, a verbal dimension, that is the words in the dialogue/monologue, and a non-verbal dimension, consisting of background noises, sound effects, and recorded music or soundtrack. Delabastita (1989, 101) has best emphasized the four basic elements that inter-relate in order to form the inter-semiotic texture of the audiovisual text: the acoustic-verbal elements, the acoustic-non-verbal elements, the visual non-verbal elements and the visual-verbal elements.

11.3 Types of Audiovisual Translation

The journey of the audiovisual text from the source language to the target language is possible in many ways, resulting in different translation modes or types of audiovisual translation. The transfer of the spoken dialogue of an original audiovisual programme into another language can be dealt with through two basic approaches, resulting in three main translation modes. The oral input can be either transformed into written output, the process being known as subtitling, or it can remain oral, like in the original production, a case in which, the original source language soundtrack is replaced by a target language soundtrack, the process being known as revoicing. When the replacement is total, the target viewer no longer has access to the original soundtrack, and the resulting translation mode is known as dubbing or lip-sync. When the replacement is only partial, the original dialogue remains faintly audible in the background, and the resulting translation mode is known as voiceover. Yves Gambier (2003, 172) divides the types of AVT in *dominant types* (interlingual subtitling or open caption, dubbing, consecutive interpreting, simultaneous interpreting, voice-over, free commentary, simultaneous or sight translation, multilingual production, such as double versions or remakes) and *challenging types* (scenario/script translation, intralingual subtitling or closed caption, with variations of subtitling for the deaf and hard-of-hearing, live or real-time subtitling, audio description for the blind and visually impaired).

Regardless of the type of AVT s/he works with, the translator deals with the whole triadic structure of the AVT (image/words/sounds). This is to say that even if s/he does not actually translate or incorporate the extratextual elements in the translation, his/her choices are influenced by these elements.

11.4 Extratextual Elements in Subtitling

Subtitles are governed by a series of technical, time and space constraints 'to the point of making language hostage to parameters'. (Neeves, 2009, 150). According to Karamitroglou (1998), the standard parameters that dictate the subtitler's options are related to: space (position on the screen, number of lines, text positioning, number of characters per line, typeface and distribution, font colour and background), time (duration of a full two-line subtitle, duration of a full single-line subtitle, duration of a single word subtitle), punctuation and letter case and target text editing.

Despite being an addition to the original audiovisual product, subtitles are most often analyzed in terms of loss. The quantitative loss is an acceptable logical consequence of the constraints enumerated earlier. The transition from the spoken code to the written one can only be done by reducing the dialogue to the minimal function of conveying the plot. In order to obtain effective subtitles, the translator has to eliminate and simplify all 'disposable' elements that would not alter the meaning of the original dialogue. It is estimated that the original dialogue tends to be reduced by between 40% and 75% (Chiaro, 2009, 148) in order to give the audience a chance of watching the programme effortlessly, that is of reading the subtitles without losing track of what actually happens on the screen and without realizing that they make a conscious effort to follow the captions.

The three aspects considered earlier, namely the time and space constraints governing the subtitling process, the fact that these constraints lead to condensation and the necessity of taking into account extratextual markers face the translator with a challenging paradox: s/he has to obtain a written target text that is a shorter version of a longer spoken text by actually incorporating extratextual markers in the translation, while rendering the same message and creating a similar effect upon the target audience.

Extratextual elements actually represent all the elements of the audiovisual text, except for the spoken dialogue/monologue: visual non-verbal, acoustic non-verbal and visual verbal elements. There are two main issues to be addressed in this respect: which of these elements are prone to translation and to what extent do extratextual markers both influence translation and determine the translator's choices?

Different answers are valid for different types of extratextual markers.

Non-verbal elements belonging to the visual code (namely actor's movements, gestures, and facial expressions, scenery, costumes, lightning, and the like) obviously need no translation. They might nevertheless carry a heavy load of culture-bound references, yet they remain self-explanatory. The way they are processed by the target audience depends mainly on the very profile and cultural background of the audience itself.

Non-verbal elements belonging to the acoustic code (namely background noise, sound effects and recorded music) do not necessarily need to be rendered through translation. In contrast to speech and paralinguistic features, sound and music usually need no translation. Ella Shohat and Robert Stam (1985, 37) offer a broader perspective on the concept of 'non-verbal', explaining that language ultimately pervades both the music and the noise-tracks, which can embrace, at least by association, linguistic elements. Recorded music is either accompanied by lyrics or evokes lyrics in the target viewer's mind, being 'deeply embedded in social discourses, including verbal discourses'. Recorded noises are not 'necessarily "innocent" of language' themselves, and 'the image track itself is infiltrated by the ubiquitous agency of language'. The same authors point out the fact that 'camera angles can literalise specific locutions such as "look up to" or "oversee" or "look down on"'. Joselia Neeves (2009, 153) asserts that 'viewers have grown to understand filmic conventions and have come to associate musical types with certain genres and with particular filmic effects'.

This kind of association is so strong sometimes, that certain music tracks or famous songs acquire intratextual dimensions. There is a famous scene in *Forrest Gump*, where Forrest is interviewed about his experience in China and John Lennon asks him questions that are actual lyrics from one of his most famous songs, *Imagine*:

Dick Cavett: *You had quite a trip. Can you, uh, tell us, uh, what was China like?*
Forrest: *Well, in the land of China, people have got nothing at all.*
John Lennon: ***No possessions?***
Forrest: *And in China, they never go to church.*
John Lennon: ***No religion, too?***
Dick Cavett: *Oh, hard to* imagine.
John Lennon: *Well, **it's easy if you try**, Dick.*

Example 1. Dialogue excerpt from Forrest Gump

Unless the lyrics are rendered in English in subtitles, the humorous effect along with the intercultural reference effect are completely lost. When choosing a domesticating approach, the subtitler has to be careful not to go too far as there is a risk of contradicting the general intention and nature of the film itself. *Forrest Gump* is a film about the American Society, with its 20th century history and symbols. Domestication of such symbols through translation does not really make much sense, as besides technical and spatial constraints imposed by the subtitling process, the linguistic and cultural transfer is subject to demands of relevance and adequacy.

Recorded music can be dealt with in various ways in subtitles, depending on the programme genre, the profile of the audience and so on. For instance, in subtitled animation movies, the lyrics of the songs are nearly always translated in subtitles. This is firstly because kids are less likely to grasp the meaning of the song from the soundtrack or because the lyrics are connected to the 'action' of the movie. In animation movies songs are not interrupted; it is usually one of the main characters who performs the singing. There is no additional 'action' taking place on the screen and no dialogue is superimposed on the lyrics. In other types of movies, even if the song is meant to reinforce the action or to transpose the audience into a certain kind of mood, or even to produce, through contrast, a comic effect, song lyrics cannot usually be incorporated in the captions because the song is heard in the background of the dialogue.

Another element of acoustic non-verbal nature, which can pose numerous challenges to the translator is canned laughter. It is mainly specific to sitcoms and stand-up comedy and represents the feedback of the studio or live audience for the comic effect produced by the original dialogue. It obviously does not make the subject of translation, yet it greatly influences the translator's decisions. Because normally the sound of laughter triggers more laughter, canned laughter might be a helpful tool for the subtitler since it signals the audience the presence of a funny element. At the same time, it adds extra pressure for the subtitler, who has to make sure that the perlocutionary effect was transferred in the target text with surgical precision so that the canned laughter instance follows immediately after the respective caption; otherwise the whole comic effect would be compromised and the final audience would feel frustrated when they hear the studio audience laugh at a joke they did not have a chance to get. The text is literally dynamic and the target viewer has no second chance to re-evaluate a joke. In addition, canned laughter sequences might also vary in length, signalling, thus, the degree of 'funniness' of a particular joke or a humorous instance. Regardless of the fact that some words or phrases may be more jocular in one language than in the other, the subtitler has to make sure that the effect his/her translation of the joke has on the target audience is of the same intensity the original joke had on the studio audience. Whenever the subtitler fails to render the humorous effect in the target language, the target viewer who has no or poor command of the source language feels frustrated, perceiving the situation as a case of mistranslation. Moreover, the translator cannot remain 'invisible', mainly due to the fact that the target audience has permanent access to the original dialogue, and the subtitles are under the permanent scrutiny of those members of the target audience who have a certain command of the source language.

Verbal elements belonging to the visual code (signposts, street signs, banners, newspapers, letters, notes, etc.) need either be translated as individual items or incorporated in the translation of the original dialogue proper. A part of these elements may complement the dialogue with valuable information. The subtitler can insert them in separate captions using appropriate editing techniques (brackets, italics, different font size or colour). Other visual verbal elements may need no translation at all (banners, street signs) as they are part of the scenery and the viewer can instinctively decode/process the information they contain. However, there are instances when ignoring visual nonverbal elements faces the subtitler with the risk of being discredited in front of the target audience. Visual non-verbal elements are usually subtitled in cartoons and animation movies, where their omission could prevent the target audience from following the action.

Fawlty Towers provide such an example of a recurrent extratextual element which is supposed to enhance the humorous effect. The opening shot of each episode shows a sign, which initially reads 'Fawlty Towers'. Throughout the first series, letters go missing or are pulled askew. Variations of these anagrams include: 'Farty Tower', 'Warty Towels', 'Watery Fowls', 'Flay Otters', 'Fatty Owls', 'Flowery Twats', and 'Farty Towels'. Left untranslated or not even hinted at in subtitles, these elements might bring a feeling of frustration upon the viewer who lacks a good command of English.

Unfortunately, either because of habit or because of the restrictive rules governing the process of subtitling 'film translators tend to be vococentric, concentrating on spoken dialogue' (Ella Shohat & Robert Stam, 1985, 47) and leaving certain ironies and nuances to be perceived only by the viewers who are familiar enough with the source language.

11.4.1 The Target audience – extratextual parameter influencing translation strategies

There are elements outside the text, which greatly influence the subtitler's choice in terms of translation strategies. The profile of the target audience is one of the parameters dictating how interventional a strategy should be in rendering a culture-bound reference, for instance, so that the target viewer does neither feel patronised or overburdened by the target text. In the particular case of audiovisual translation, the translator's choices are conditioned, motivated, encouraged, or restricted by the target audience from a double perspective: first of all, the profile of the target audience is one of the most important parameters a subtitler has to consider when making an active translation choice, and secondly, the target audience is constantly exposed to the original source text, which makes the translation 'vulnerable'. In Diaz Cintas' terms (2010, 344) subtitling is 'a supplement to the original programme, which (...) remains intact in the target culture for all to watch and to hear'.

The audiovisual text is, paradoxically, more than a text (as it incorporates several codes and extratextual elements) and less than a text at the same time (as it is constrained and restricted by its technicality and the policies of the broadcaster, while being shaped and adapted to the profile of the audience). While a whole range of elements and factors add to the meaning of the AVT (reinforcing and contradicting the message of the spoken dialogue), another series of factors, some depending on the translator and some way beyond his/her reach cut out from the body of the text, fragment and even distort it. This is precisely why any taxonomy of translation strategies that apply to the AVT (as deduced by researchers comparing the source language text with its final subtitled target text) needs to be validated by a secondary process, in which all the co-textual, para-textual, contextual and pragmatic parameters influencing the translation process should be carefully weighted.

Some of these parameters are related to the specific nature of the text, some others are connected to the technical nature of subtitling, some are skopos-related, while some are simply pragmatic matters.

In order for the target viewer to effortlessly watch and enjoy a subtitled product, the subtitler (and team of technicians backing up/supporting/amending the target text produced through translation) must have a deep knowledge of the target audience's profile. Elements as diverse as the age range, educational and cultural background, previous exposure to the same type of programme, social status, and so forth should be carefully considered. As well as this, pragmatic matters related to deadlines and financial remuneration of subtitlers are responsible for the quality of their translations and their ability to make full use of the time consuming interventional strategies.

The profile of the target viewer is of great importance whenever culture-bound references are at stake. According to Pedersen (2005, 2), an Extralinguistic Culture-bound Reference (ECR) is a 'reference that is attempted by means of any culture-bound linguistic expression, which refers to an extralinguistic entity or process, and which is assumed to have a discourse referent that is identifiable to a relevant audience as this referent is within the encyclopedic knowledge of this audience'. The list of strategies proposed by Pedersen apply to what he calls 'translation crisis points' (puns, poetry, quotations, allusions), which correspond to Leppihalme's 'culture bumps' (Pedersen, 2005, 2). On a Venutian scale, the strategies range from the most foreignizing to the most domesticating ones, although Pedersen prefers using the more neutral terms *source-language oriented and target-language oriented.*

Although retention is by far the most common solution for a translation crisis point, its success depends on which level of transculturality it is placed. Pedersen (2005:10) makes a very useful distinction between three methodological relevant levels of transculturality, explaining that 'the degree of transculturality of an ECR deals with how familiar it is to the ST and TT audiences'. According to these distinctions, an ECR can be transcultural (i.e., not bound to the source culture, but retrievable from the encyclopaedic knowledge of both ST and TT audiences), monocultural (bound to the source culture and less identifiable to the majority of the TT audience; these ECRs represent translation crisis points) and microcultural (bound to the source culture, but too specialized or local to be within the encyclopaedic knowledge or both the ST and TT audiences).

The profile of the target audience is equally important in the subtitler's decision regarding the best strategies to be used when dealing with taboo language. Comparative studies in translation have constantly asserted that taboo language is undertranslated. However, some of these prerogatives should be revisited and brought closer to one of the questions initially raised by this research: What do subtitlers actually translate? Labelling a translation solution for a term or phrase as undertranslation immediately places it on a scale where a direct translation procedure or perfect linguistic equivalent would be the best (more or less feasible) option. Why would source taboo language be undertranslated in subtitling? Could it be because Romanian language, for instance, is less creative or productive in this field? Alas, no, on the contrary, it is quite rich, both vocabulary and grammar-wise, and it has been so for centuries. A more down-to-earth and widely accepted explanation is that the written word is more powerful than the uttered one. The target viewer might feel offended or burdened by captions containing perfect dictionary equivalents of swearwords. Moreover, specific policies of the broadcasting companies might, for obvious reasons, exclude such techniques.

Undertranslation of taboo language and swearwords often results in omitting the problematic element altogether. There are cases in which the swearword is initially undertranslated, rendered by a 'softer' equivalent, then omitted in the captions, although it is repeated in the spoken dialogue. Numerous examples are provided by stand-up comedy routines that are extremely rich in taboo language and where the rhythm of speech is extremely high. Omission is consequently requested by the need for textual economy as well as by the fact that the viewer can actually hear the monologue and the swearwords are easily detectable, so there is actually no need to render them in the captions. They are more of a brand, a part of the comedian's persona, a verbal habit. For instance, Billy Connolly, in a 3:31 minute fragment of a routine on the Solar System uses the f-word with a frequency that could match his heart rate (which is probably quite high, considering the amount of body language involved), while Craig Ferguson utters it 52 times in a 2:14 minute fragment of his *Does This Need to Be Said?* routine, where he admits being a very 'cursy' person. The humour here is actually triggered by the fact that as the host of a TV show, he must comply with the restrictions of the TV company and refrain from using taboo language. The subtitler should actually rely on the fact that the target viewer who is likely to watch a stand-up comedy routine performed by Craig Ferguson has a specific profile and specific knowledge about the comedian and his career, as well as a fairly good command over English, and consequently uses the subtitles only for guidance, to get the more intricate jokes. Taboo language is, otherwise, easily detectable and comprehensible and does not need to be rendered in the captions.

Swearwords seem to be more problematic and resistant to translation when they appear in isolation, rather than when they are used extensively. This is mainly because the overuse of profane language diminishes the impact it has both on the SL viewer and the TL viewer, who has a certain command of the SL. When used in isolation, swearwords are meant to express a certain state of mind of the speaker, an attitude or a strong emotion. Even in this case, their rendering in the TT seems to be regarded with conspicuous caution. However, this tendency is not to be defined necessarily as undertranslation, but rather as a sort of register calibration.

Nord (2010, 123) speaks about intertextual and intratextual coherence, stressing out the fact that 'in order to make the target text work for a specific target audience, the translator should produce a text that conforms to the standard of what Vermeer terms *intratextual coherence*, which means that the addressed audience should be able to make sense of it and that the text should be acceptable for them'.

The use of taboo language is so frequent, so 'normal' in American movies, for instance, that it produces no surprise to the source viewer. On the contrary, the Romanian cinema and movie industry has rather stayed away from profane language. This is why translating it bluntly, by means of perfect equivalents, would rather sound like over-translation. When calibrating the register, the subtitler should be careful in choosing a vocabulary range that would make the character sound credible to the Romanian viewer and in placing that character on the social scale originally intended for him/her by the creators of the movie.

Analysing and comparing a whole list of taxonomies of translation strategies could, at some point, seem dull if these classifications were considered outside a range of parameters motivating (or de-motivating) the translator's and more particularly, the subtitler's choices. Lists of do's and don'ts can, of course, be drawn; errors can be identified and more felicitous solutions can always be prescribed. Yet, at least as far as ECRs are concerned, each translator's choice and active decision apply to a unique context. Although audiovisual translation has become a highly distinct branch of Translation Studies, there is a huge gap between theorists and practitioners. On one hand, globalization and intercultural communication via audiovisual products has proliferated so rapidly that the assumption that the subtitlers have the time to read the literature in the field and apply models offered by translation scholars would be naïve. Practitioners deal with some very down-to-earth issues, such as tight deadlines, constraints, policies and censorship imposed by the broadcasters and producers, rapid changes in technology, remuneration, contracts and (repetition intended) tight deadlines. On the other hand, globalization comes in handy when ECRs are at stake. An ECR that was monocultural a few years ago, might easily have become transcultural in the meantime (thus requesting a 'less' interventional strategy). The type of audiovisual text being translated matters equally. A sitcom, for instance, addresses a specific-profile audience. A pact is concluded from the very beginning: the target viewer watches the series in order to be entertained, s/he is expecting the punch-line, which triggers laughter being backed up by verbal and non-verbal extra-textual elements, such as character's gestures, facial expressions, canned laughter and the like. The target viewer is also 'trained' by his/her previous experience (previous episodes of the series, other encounters with sitcoms or comedy movies). This is to say that s/he might have come across the same ECR, or that in decoding an ECR, s/he does not only rely on his/her encyclopaedic knowledge.

In the particular case of subtitling humour, taxonomies cannot be applied prescriptively, as subtitling deals with an extremely complex type of text and each 'translation crisis point' challenges the subtitler in a unique way. Although restrictive by nature, the process of subtitling can incorporate the whole range of translation strategies described by scholars, from the most domesticating to the most foreignizing ones.

11.4.2 Extratextual elements in other forms of audiovisual translation: SDH and AD

Whether or not to incorporate extratextual elements in audiovisual translation is ultimately the translator's choice, and the average viewer might judge this choice in terms of how successful the translation is. For some categories of target viewers, though, the signalling of extratextual elements in translation is actually a necessity. Accessibility is a key concept in AVT and is globally encouraged for more or less commercial purposes by producers of audiovisual programmes. It is expected that the needs of groups, such as the deaf, the hard-of-hearing, the blind and the visually impaired should be catered for with more consideration. Fortunately, this is actually the case in Western Europe and most English-speaking countries. Consequently, accessibility to the media for people with sensory impairments has strengthened relatively new fields of expertise within audiovisual translation, namely subtitling for the deaf and hard-of-hearing (SDH) and audio description for the blind and the partially sighted (AD).

Statistically, between 1% and 5% of the population of any country are deaf or hearing-impaired. In subtitling for the deaf and hard-of-hearing (SDH), also known as (closed) captioning in American English, the actors' dialogue is converted into written speech, which is presented in subtitles of up to three, or occasionally four, lines. Subtitles generally change colour on the screen indicating, thus, the person who is talking or the emphasis given to certain words within the same subtitle. Besides the dialogue, they also incorporate all paralinguistic information that contributes to the development of the plot or to the creation of atmosphere, which a deaf person cannot access from the soundtrack, e.g., a telephone ringing, laughter, applause, a knock on the door, and the like.

In order to tackle down the typology of this particular AVT mode, Joselia Neeves (2009, 150) draws attention upon the importance of being aware, as a translator, of the specific needs of the audience: 'subtitles are all about reading (...) to the Deaf and HoH subtitles are essential, rather than redundant. They are the visual face of sound. For the HoH they are a stimulus and a memory exercise; for the Deaf, they are the only means to gain access to aural information. However redundant, sound and image tell different stories'.

On the Romanian audiovisual scene, this type of translation is poorly represented, although the deaf community is quite large. However, a Law Project regarding Technical and Social Assistance for the Deaf and Hard-of-Hearing issued by the Romanian Parliament in October 2013 states that 'The Romanian Public Television shall provide subtitles for 80% of all cultural, political and general interest programmes broadcast every year'.

Visual impairment is one of the most age-related disabilities and the vast majority of blind and partially sighted people are elderly. As demographic trends show that the number of elderly people is on the increase, one can only expect that the percentage of people with visual impairment will also increase in the near future.

Cintas (2008, 2) describes AD as 'an additional narration that fits in the silences between dialogue and describes action, body language, facial expressions and anything that will help people with visual impairment follow what is happening on screen or on stage'. The procedure is again poorly represented in Romania, one of the reasons being that we are one of the countries where subtitling is favoured and the audio description of a foreign audiovisual product would only make sense if that product were dubbed. This apparent impossibility could be overcome by means of audio subtitling in order to make subtitled programmes accessible to the blind in countries where a large percentage of the programmes are commercialized in a foreign language with subtitles. The website of the Romanian Association for the Blind (ANR) indicates that the association has about 80,000 members. Yet, apart from information regarding a recording studio for audio books, that was founded in 1966 and has produced over 5,000 book recordings, there is no mention about any sort of audiovisual programme. On the other hand, a few hundred kilometres away, EBU organizes this year the 5th Advanced Seminar on Audio Description (ARSAD), demonstrating the increasing concern for the development of AD in Europe. This is, however, a social matter and we can only hope that in the nearest future, Romanian researchers in the field of AVT will be able to come up with their own examples and data about AD in Romania.

11.5 Conclusions

The extratextual elements give the source language text its full meaning, which can be completely altered if they are neglected in translation. Subtitling is seen as a vulnerable form of translation. What is said and implicitly translated is only a part of the message. The way it is said (sound, intonation) together with visual and auditory markers (visual signs, gestures, postures, editing techniques, soundtrack) is another part. Thus the reader is also the 'viewer' and 'hearer' of a polysemiotic entity that somehow imposes itself on its receptor with a certain speed, range of images and sounds. Since what we call non-verbal or extratextual markers are still marked by the presence of language, one might legitimately wonder if the subtitler, already challenged by the economy of space and time characteristic to subtitling, could find any resources and/or strategies to incorporate the meaning of these language-associated nonverbal elements into his/her translation. Due to the complex nature of the audiovisual texts, no 'rules' can be established regarding the signalling, description, translation or adaptation of extratextual elements in the captions. Nevertheless, the subtitler should be fully aware of the profile of the target audience in terms of age, cultural background, sensory impairments, etc. in order to make sure that the perlocutionary effect of the original text is achieved without the risk of either over- or under-stressing the target viewers.

Acknowledgement: This work was supported by the strategic grant POSDRU/159/1.5/S/133652 and co-financed by the European Social Fund within the Sectorial Operational Program Human Resources Development 2007–2013.

References

Chiaro, D. (2009). Issues in Audiovisual Translation. In J. Munday (Ed.) The Routledge Companion to Translation Studies, (pp.141-165). London/New York: Routledge.

Delabastita, D. (1989). Translation and Mass-communication: Film and T.V. Translation as Evidence of Cultural Dynamics. Babel 35(4), 193-218.

Diaz-Cintas, J. (2008). Audiovisual Translation Comes of Age. In D. Chiaro, C. Heiss, & C. Bucaria (Eds.) Between Text and Image. Updating Research in Screen Translation, Benjamin Translation Library, Volume 78, (pp. 1-11)

Diaz-Cintas, J. (2010). Subtitling. In Y. Gambier & L. Van Doorslaer (Eds.) Handbook of Translation Studies, John Benjamins Publishing Company, Volume 1, (pp.344-348)

Gambier, Y. (2003). Introduction – Screen Transadaptation: Perception and Reception. Screen Translation. The Translator, Volume 9, no. 2, (pp. 171-189). St. Jerome Publishing, Mancester, UK.

Karamitroglou, F. (1998). A Proposed Set of Subtitling Standards in Europe. Translation Journal,Volume 2 (retrieved on January, 6, 2015, from http://accurapid.com/journal/04stndrd.html)

Neeves, J. (2009). Interlingual Subtitling for the Deaf and Hard-of-Hearing. In J. Diaz Cintas & G. Anderman (Eds.) Audiovisual Translation – Language Transfer on Screen, (pp.151-170). Palgrave Macmillan

Nord, C. (2010). Functionalist Approaches. In Y. Gambier & L. Van Doorslaer (Eds.) Handbook of Translation Studies, John Benjamins Publishing Company, Volume 1, (pp. 120-124).

Pedersen, J. (2005). How is culture rendered in subtitles? (retrieved on August, 2, 2015, from http://www.euroconferences.info/proceedings/2005_Proceedings/2005_Pedersen_Jan.pdf)

Shohat, E. & R. Stam. (1985). The Cinema after Babel: Language, Difference, Power. In Screen 26 (3–4), 41. doi:10.1093/screen/26.3-4.35 (retrieved on January, 22, 2015)

Elena Laura Vulpoiu

12 Subtitling in Romania and Spain: A Contrastive Analysis

12.1 Introduction

It is a part of our daily lives, but apparently, it does not receive the attention it deserves, except from theorists and translators. Audiovisual translation is an activity that by means of film or television series changes our feelings, entertains and educates children with cartoons, or informs and instructs us with documentaries of all kinds.

The importance of audiovisual translation in society is fundamental, regardless the country or the method chosen to transmit the contents of foreign programes, with dubbing and subtitling as the most common forms. They are practiced all over the world.

It's difficult to decide which is the best method for transferring an audiovisual text, either dubbing or subtitling. On one hand, in many cases, the decisive factor for the election is not quality but cost. On the other hand, the social class to which the viewer belongs has an influence on the public's preference for one or the other modality. That is why on the basis of political, cultural, economic or social reasons; each country chooses one of the two forms of translation mentioned above. However, in most countries both modes are used.

Theorists and translators also have their preferences and have provided reasons justifying one form or another. This has caused a historical clash between the two techniques because there are some academics and professionals who advocate one, but reject the other.

I will not get into this debate, although at the beginning of my studies, without knowing the theory of audiovisual translation, I had chosen the subtitling. At present, I am aware that both, subtitling and dubbing, have different functions and they meet the requirements of different groups of spectators. It is true, however, that many times the choice is taken considering the price and not the quality of the product that is offered to the viewer.

The motivation for this investigation stems from the desire of understanding the practice of audiovisual translation mostly used in my country, Romania. It is the subtitling method. There are several monographs and studies on subtitling practices based on countries in Western Europe, but very few on this practice in other countries, especially in East Europe. This is the first empirical study which presents results about the present process of subtitling in Romania.

12.2 Subtitling in Romania

Romania is one of the countries of former Eastern Europe, which is making use of subtitling as an audiovisual translation method, for both films and television. Whether for political or economic reasons, this type of translation has been mostly used as a tool to translate the avalanche of programs and films that have been made in the audiovisual field after the revolution of 1989. Although only a few years ago, in 2011, the deputy, Victor Socaciu, has proposed in parliament to change subtitling for dubbing as a way of protecting the national identity; his proposal was rejected by a large majority. The followers of subtitling insist on asserting that a film is conceived to be seen in its original language. They argue that emotions are transmitted to the viewer through the voice and intonation. Emotions cannot be transferred similarly via dubbing. Besides, professionals of this field, like Irina Nistor, claim that this change cannot take place suddenly because the public is already habituated to subtitling. Another factor is the fact that broadcasters are not prepared to deal with this change which involves significant costs, mainly by the need for specific training of professionals in the process of dubbing (Ziare.com, 2011).

In the book *Cine y Traducción*, Federico Chaume (2004) defines this modality in the following way:

> Subtitling consists in incorporating a written text in the target language to the screen, where a film is shown in its original version, so that these subtitles coincide approximately with the interventions of the actors on the screen (Chaume, 2004:33). [my translation]

According to this definition, the foreign product to be subtitled requires considering three key components in this process: the oral word, image and subtitles. The set of these three elements, reading ability of the viewer and the screen dimensions define the basic characteristics of this type of translation. Thus, subtitles must be synchronized with the image and stay on the screen long enough so that, the viewer can read them.

Henrik Gottllieb (1997) is one of the first researchers who worked with subtitling and its procedures. In his book, *Subtitles, Translation and Idioms,* he contemplates subtitling from different perspectives, as for example history, experiments on the reaction of the viewer or translation of idiomatic expressions.

Limited by the space-time factors that the audiovisual media impose, the translation in this field is also subject to restrictions of various kinds. Furthermore, the subtitling involves a change of mode of discourse, oral to written, which usually forces the translator to omit elements of the original message. Many theoreticians understand the subtitling as an adaptation and not as a translation of the original text to the target text. But, it is imperative to take into account that, in this type of translation, delivered message is predetermined by the spatial and temporal synchrony, which imposes certain limitations on the translator. In this case, the assertion of Díaz Cintas is very suggestive:

> It is necessary to understand translation from a more flexible and diverse, less static perspective that accommodates a wide range of empirical realities, and take into account the changing nature of this practice (Cintas, 2003: 34). [my translation]

All these elements have to be well synchronised with the translation, subtitling or dubbing, to obtain quality of the final result.

Subtitling can be a challenge for a translator since the juxtaposition of the original text with the translated text may allow the viewer to make the comparison of both messages (Mayoral, 1993). This simultaneity of the two linguistic codes has consequences in the translated program or movie. However, the translator must remember that subtitling is a form of linguistic and cultural transfer, which operates at two levels simultaneously: the change from one language to another and the change from oral code to the written code (Gambier, 1996:10). In this situation, the translator must be aware of both aspects in order to offer the public an acceptable translation.

The strategy usually used by the translator is to transfer the terms of the original subtitles that conserve a close phonetic and etymological connection in both languages and can be easily recognized by the viewer when listening to the original dialogues. Thus, subtitling becomes a kind of 'vulnerable translation', as it was called by Díaz Cintas (2001), because it is exposed to criticism of the viewers who are often questioning the translator's work.

Subtitling was practiced in Romania before 1989 due to sociopolitical and economic reasons. The dictatorship of the pro-Soviet regime and the situation of political isolation in the European market prevented the influx of foreign audiovisual products, and the few that penetrated the Romanian market were subtitled.

At present, although the Romanian audiovisual market is open, the habits acquired by the public and the economic precariousness of some TV stations have determined the existence of subtitling as a common method. The method represents a lower cost than dubbing. Just a single translator is needed.

Exceptionally, a linguistic reviewer is hired for correcting the translation. Television networks contact and hire professional translators with experience, since the product will always be translated and is going to have a wide social diffusion. This implies a greater responsibility for the participants in this process, for the entire chain, and therefore, in the end also for the translator.

To shed some light on the strategies and techniques used by subtitlers in Romania and in order to get to know the guidelines and conventions that are applied in the process of subtitling in this country, I have contacted a professional translator, Mr. Bogdan Stănescu. With his support, I will try to bring to light some unknown information well kept by the television networks, since Mr. Bogdan Stănescu was the only one who was so kind to answer my questions.

12.3 Conventions of Subtitling in Romania

As a starting point for their work, translators receive from the company that hired the – in our case, the television network – the dialogue list for the subtitling of foreign programs. The translator's work begins at that moment, and she/he has to consider not only the words, but also the form of the subtitles, spelling conventions, the synthesis of information, rules of subtitling, such as reading speed, the whole set of unknown elements by the viewer, which require a perfect synchronization with the time of issuance of the dialogues for the translation of the text.

12.3.1 Form of subtitle and partition lines and subtitles

In Romania, subtitling is typically performed on two lines positioned in the lower part and centered on the screen. The maximum number of characters for each line used is 40.

When segmenting the subtitles, the translator has to note that these are, in themselves, logical information units. It is preferable to use subtitles of two lines that do not exceed 40 characters. That is easiest for the viewer. According to Stănescu, it is easier to read two short lines, than one line of 40 characters.

12.3.2 Orthotypographic conventions

The orthotypographic signs have the same use in almost all languages, but in the case of subtitling, the translators apply the conventions of each country in their translations (Cintas, 2003). Here are summarized some features that are granted to those ortotypographic signs by the person responsible for translating foreign films on the TVR1 national channel.

The comma and period are used for grammatical purpose in the target language, while the semicolon is never used in subtitles. The ellipses occupy three characters, and they are used to mark a pause or unfinished sentences. The dash is used to indicate two interventions in the same subtitle, always using the short dash. The type of quotation marks that are used are the double (‘ ‘) and they are used to appoint famous quotes, titles and foreign words. Capital letters are used to mark titles, while italics are always used to indicate the narrator's voice text. The numbers from 1 to 10 are written in letters while they are written in numbers from 11: for room numbers, addresses, days of the month, year, exact times, and measures. The point between numbers is used to mark thousands, millions and years. The abbreviations are used to mark units of measure, e.g., hours and if they appear in the original text, they are generally translated.

12.3.3 Synthesis of information: What is deleted from the original text?

According to Gottlieb, time and space are the key elements of subtitling: 'the famous and infamous time and space constraints of subtitling [...] This normally implies some measure of condensation of the original dialogue, something that is often not expected in translated texts (Gottlieb, 2004:219)'. Timing has to be perfect, which implies that the translator is forced to resort to omission and reduce information in relation to the original text. Interjections are always deleted, i.e., they are never written in subtitles. Other elements that are sometimes suppressed, depending on the situation, are the vocative, adverbs and adjectives or words well-known to the audience, but if the message is abbreviated in the target language, the original referents should not be altered. In the case of repetitions, the translator can suppress a denial, for example, the replication: 'No, we do not leave' or, as another example, ' No, no and no'; to translate this replica, it could be used as an equivalence 'On no account ' or ' No case ', always according to Stănescu. The first and last names of the characters, though they sometimes are omitted in the target text, they necessarily appear at the beginning, middle and end of the film so that if the viewer that has not seen the movie from the beginning he/she can identify characters by their name. Other sentential elements that sometimes can be suppressed are songs. Their translation has to be done as long as the lyrics is part of the plot of the movie. Generally speaking, about 40% of the original text can disappear. This figure may vary depending on the textual variety (Cintas, 2003:202). Condensation or conciseness requires prioritizing short words as they occupy less space on the screen. The translator has to know the list of synonyms and give the precise suitable word in order to prevent a distortion of the meaning or the register of the original sentence.

12.3.4 Spotting

Time and image are very important in subtitling; the viewer needs time to read the subtitles, but also to follow the thread of the film and the performances of the actors. The viewer's attention must be shared among all these channels and codes. The maximum time needed for the subtitle of two lines remaining on the screen is 7 seconds and the minimum is 5 seconds (Cintas, 2003). When it comes to subtitles that have one or two words, the minimum time is 1 second. Temporal synchrony must be perfect among the entry of a subtitle and the beginning of its enunciation, but, our interviewee added: 'We have to take into account that subtitling is performed manually and it also depends on the flair and professionalism of the person who is performing it'. Respecting these basic rules, the translator has to produce subtitles that are easily understood by the viewer in the brief time that they appear on screen (Ivarsson & Carroll, 1998:74).

12.3.5 Reading speed

The reading speed varies depending on each individual; therefore, it is necessary to establish a standard time for the subtitles to remain on the screen so that the viewer has sufficient time to read them. According to our data, in Romania, it is 7 seconds for a maximum of 80 characters (two subtitles each one of 40 characters), including spaces and punctuation. The standard application to measure the speed of reading is 10 characters per second, and the optimal measure is 15 characters per second, including spaces. The estimated time for reading and understanding a subtitle that is composed of two lines of maximum 35 characters or keystrokes per line, which means a maximum of 70 characters is 6 seconds. We note that there is a difference of one second between the estimated time in Romania and the estimated time by theorists like Ivarsson (1992), Gottlieb (1997), or Cintas (2003). For this reason, the guidelines guide us in our attempt to carry out consistent work, but nevertheless, they are open to changes and alterations (Cintas, 2003:119).

Through the information provided by Bogdan Stănescu, our interviewed translator, we have been able to determine in detail the guidelines and conventions of subtitles that are used in this country, especially in the TVR1 channel. We note that translators adapt their work to the conventions, although sometimes, either by a structural difference between the languages utilised or cultural restrictions, they are forced to use the techniques and methods that they consider convenient for each particular case.

12.3.6 Use of software

To perform the subtitle, the national television TVR1 uses a non-commercial program created especially for this institution.

12.3.7 Paratextual Elements

The delivery of the subtitle document is done in a text document (Word, Open Office), which includes the name and surname of the translator, the title of the work in original version and subtitled, and in the case where foreign elements are added to subtitling these are placed at the end of the document.

12.4 Comparison with the Conventions in Spain

In Spain, the mode used predominantly for the translation of foreign products is dubbing, nevertheless, some programs are broadcasted on television that also use subtitling. Despite the heterogeneity observed in audiovisual products you find in the market, the realization of subtitles is not random. It is a process that has to adhere to conventions that the translators have to keep in mind when doing their work. It can be for the simple reason that many broadcasters, distributors, or recording studios lack the style books that are offered to the translator as guidelines for submitting of the subtitles.

Generally the orthotypographic rules are similar for all countries, but their use is sometimes particularized in subtitling and the subtitles are configured depending on the foreign product that is translated into the target language, or depending on the translator who does the work.

The most updated exhaustive work on subtitling, and subtitling and conventions in Spain in particular, is the book *Teoría y práctica de la subtitulación*, by Díaz Cintas (2003), whose analysis will be used to present the characteristics of subtitles in Spain.

12.4.1 Form of subtitle and partition lines and subtitles

In Spain, the subtitles are presented in the same way as in Romania, using two lines centered in the lower part of the screen, with a variable number of 35 to 37 characters per line. Their segmentation is performed taking into account the units of logical information in which the target text is structured.

12.4.2 Orthotypographic conventions

In Romanian interrogative, and exclamatory sentences are marked with the appropriate sign at the end of the respective sentences, whereas in Spain sentences of this type are marked with conventional signs, both at the beginning and end of the sentence. That leads to the consumption of more screen space in Spanish. Díaz Cintas claims that no punctuation that fulfils a specific function should be omitted since it is a feature of the target language rules, but the subtitles should not be overburdened with punctuation marks that do not provide anything new and can hinder the reading by the viewer (Cintas, 2003:162). The accent is another orthotypographic sign that does not exist in the Romanian language, but in Castilian their use is mandatory both lowercase and uppercase, when appropriate, since the spelling of the subtitle has to be normative. Regarding the other punctuation marks, when comparing the conventions in Romania with the conventions in Spain, we observe the same use in Spanish and Romanian: for example, both languages try to avoid using the semicolon because it can confuse the viewer, since it is easy to associate this type of sign to a point and to consider the sentence finished.

The short script is often used to mark shifts of dialogue, the double quotes to mark appointments, invented expressions, literature and literary references, or to mark the terms that are borrowed from another language. Italics are used to mark the voiceovers that come from people who are not on the scene, but it can be heard on radios, stereos, etc. The uppercase letter is currently used to mark the titles or the names. Unlike Romania, in Spain, long numbers like millions and billions are not transcribed with numbers, but with the equivalent words (million, billion) to avoid difficult reading to the viewer. When performing a comparison between the two models of subtitling: Romanian and Spanish, we can see that some orthotypographic conventions are different and vary in each country.

12.4.3 Synthesis of information: What is deleted from the original text?

The strict time-image relationship requires from the translator to make a synthesis of information to configure the subtitles. Thus, the reduction can be partial (condensation/conciseness) or total (elimination/default/deletion), but always bearing in mind the principle of relevance, and not omitting information that may be essential to the understanding of the history. Díaz Cintas affirms that, in a general mode, 40% of the original text will not disappear, and that this number may vary depending on the textual variety. It may seem that a large amount of information is suppressed, but this technique it is applied:

> [...] a large number of repetitions, exclamations, speeches parasites or words that play a phatic function and whose 'loss' does not have a negative effect on the mediation of semantic information (Cintas, 2003:203). [my translation]

Condensation or conciseness is possible by prioritizing short words, as opposed to long words since those would occupy more space on the screen. The translator has to know how to use synonyms adequately for not distort the meaning or the registration of the original sentence. Every omission is a decision that can affect comprehension. That is why, the translator has to assess whether the viewer can access all information that is transmitted without too much effort. The elements that are usually ignored are phrases or expressions that are repeated in the original language. Sometimes, expressions that are easily understood are added by similarity in both languages. That is, references to people or places displayed on the screen or proper names, in the moment when it is considered that the viewer knows their identity. All these techniques must be used with caution, without producing a big difference between the original products and the subtitling. The translator has a fundamental role in this process. Like with all interpretations, quality depends on the intelligence, knowledge, artistic sensibility and determination of the subtitler (Ivarsson & Carroll, 1998:85).

12.4.4 Spotting

Another great responsibility of the subtitler is the reproduction of subtitles, respecting the basic criteria of linguistic cohesion. Therefore, a translator has to give maximum attention to the textual and discursive coordinates but, in addition, to the thematic coherence and to lexical cohesion of the subtitles. While respecting these basic rules, the translator must produce subtitles, which are easily understood by the viewer, in the short time that they appear on the screen (Ivarsson & Carroll, 1998:74). For this, he/she tries not to separate the verbs from their complements with the purpose to facilitate the reading and understanding of the viewer. If a sentence is the answer to a question, the ideal is that both appear in the same subtitle so that, a temporary pronounced lapse is avoided, which can hinder the understanding of the information transmitted. The phrasal verbs are changed by simple verbs. Thus, the translator gets more screen space for subtitles. As we can see, the translators in both Romania and Spain, adjust their work to conventions, but sometimes, either by a structural differences between the languages involved or because of cultural restrictions, they are forced to employ the techniques and methods they consider appropriate in each case.

12.4.5 Reading speed

The estimated time for reading and understanding of a subtitle which is composed of two lines (with a maximum of 35 characters per line) or keystrokes per line (which means a maximum of 70 characters) is 6 seconds. We can see that there is a difference of one second between Romania and the estimated time in Spain. The time calculated for reading is not always the same; for that reason, Díaz Cintas states that

> Some companies may prefer a reading speed faster or slower over the whole program, or specific scenes and moments, and it will assign values slightly different to the duration of dialogues. We have to understand, therefore, as guidelines that guide us in our attempt to carry out a consistent labor, but simultaneously they are open to changes and alterations (Cintas, 2003:119). [my translation]

12.4.6 Use of software

Presently, in the market, there is a variety of computer programs that are used for subtitling. Among the free software programs, in Spain are used: Subtitle Workshop, Aegisub and VisualSubSync. The commercial programs, that stands out WinCaps, Spot, EZ Titles Fab and Swift (Ferriol, 2012: 39–48).

12.4.7 Paratextual elements

The delivery of the subtitled document is usually done in the file generated by the corresponding program of subtitling (Ferriol, 2012), and to a much lesser extent, in a text document (Word, Open Office), which includes the name and surname of the translator, the title of the work in original version and subtitled, as occurs in Romania.

12.5 Conclusions

By comparing subtitling conventions between Romania and Spain, this study showed that although the conventions are similar, there can be observed differences between one country and another. These differences are due to, especially, the guidelines the translators receive from the channels that transmit the product.

I take this opportunity to thank the translator Bogdan Stănescu for his unconditional collaboration. He let us immerse in the world of audiovisual translation in Romania, a discipline we consider to be essential in the field of translation in general. My sole purpose was to present some of the standards and guidelines that are used in audiovisual translation in Romania since there are very few studies on this topic.

I consider that my observations cannot be generalized, considering that we have enjoyed the cooperation of one translator, but we do not rule out other studies focusing on the same subject, encompassing more views of the Romanian professional translators.

References

Chaume, F. (2012): Audiovisual Translation. Dubbing. Manchester: St. Jerome.
Chaume, (2004): Cine y traducción. Madrid: Cátedra.
Corrius, M. (2008): Translating Multilingual Audiovisual Texts. Priorities and Restrictions. Implications and Applications. Tesis doctoral. Universitat Autònoma de Barcelona.
Díaz Cintas, J. and A. Remael (2007): Audiovisual Translation: Subtitling. Manchester: St. Jerome Publishing.
Díaz Cintas, J. (2003): Teoría y práctica de la subtitulación. Barcelona: Ariel.
Díaz Cintas, J. (2001): La traducción audiovisual: el subtitulado. Salamanca: Almanar.
Duro Moreno, M. (2001): La traducción para el doblaje y la subtitulación. Madrid: Cátedra.
Franco, E., A. Matamala and P.Orero (2010): Voice-over Translation. An Overview. Berna: Peter La Mayoral Asensio, R. (1993): La traducción cinematográfica: el subtitulado, en Sendebar, Boletín de la Facultad de Traductores e Intérpretes de Granada, No. 4, 45-68.
Gambier, Y & van Doorslaer, L. (eds). (2009): The Metalanguage of Translation. Amterdam & Philadelphia: John Benjamins.

Gambier, Y. (1996): La traduction audiovisuelle un genre nouveau?, en Gambier, Yves (ed.), 1996, Les transferts linguistiques dans les média audiovisuels, Villeneuve d'Ascq (Nord), Presses Universitaires du Septentrion, pp. 7-12.
Gottlieb, Henrik. (2004): Subtitles and international anglification. Nordic Journal of Eglish Studies: 3(1). Special Issue. World of Words. A tribute to Arne Zettersten: 219-230.
Gottlieb, H. (1997): Subtitles, Translation and Idioms. Copenhagen: Copenhagen University. Center for Translation Studies and Lexicography.
Ivarsson, J. and M. Carroll (1998): Subtitling, TransEdit HB, Simrishamn.
Ivarsson, J. (1992): Subtitling for the Media. A handbook of an art. Stockholm: Transedit.
Marti Ferriol, J.L. (2012): Nueva aproximación al cálculo de velocidades de lectura de subtítulos, Trans 16, 39-48.
Newmark, P. (1982): Approaches to Translation. Oxford: Pergamon.
Toury, G. (1995): Descriptive Translation Studies and Beyond. Ámsterdam/Filadelfia: John Benjamins.
Vermeer, H. J. (2000): Skopos and Commission in Translational Action, en Venuti, Lawrence (ed.) The Translation Studies Reader (pp 221-232), New York: Routledge.

Webgraphy

Vreti filme dublate sau subtitrate? - Sondaj Ziare.com. Avaible in: http://www.ziare.com/media/televiziune/vreti-filme-dublate-sau-subtitrate-sondaj-ziare-com-1092879

Part III: **Language Teaching and Learning in the Age of Technology**

László Imre Komlósi

13 Digital Literacy and the Challenges in Digital Technologies for Learning

13.1 The Digital Age: A paradigm Change in Social Cognition

The digital age has brought not only new ways of collecting and processing information for human beings (and other intelligent, reasoning entities as well) in the different societies on the globe – regardless of the social and political order or ideological commitment – but it has also made it necessary for researchers to develop new ways of thinking about and understanding digital culture in the quest for answers to complex questions about literacy and human learning and social cognition.

In the present study, I make an attemp at summarizing some of the converging findings in my research in the fields of cognitive linguistics and cognitive anthropology, which I have conducted in the course of the past decade. My interest has evolved around research on metaphorical and metonymical meaning structures together with meaning extension via conceptual integration, the theory of rationality and mental state attribution, discursive reasoning and argumentative skills development under the conditions of interactive, contextualized verbal behavior. I intend to survey briefly the reciprocal relationship between verbal behavior and social cognition in which both cognitive and affective processes play a crucial role. I plan to show how certain learning techniques get amplified, while others decrease in importance due to the changing nature of accessing and processing information in the digital era.

The paper is based on the observation that unprecedented development and innovation in information and communication technologies have exerted unforeseen impact on social cognition, information processing and human learning. We have witnessed the appearance of smart systems of various sorts whose interactive elements include reasoning (i.e., reasonably computing) entities with both human and non-human properties. It is realistic to claim that learning and knowledge management in digital social space are bound to take place and consist in the interaction between intelligent cognizing entities, human and non-human alike.

The paper identifies the observable causes of a paradigm change in social cognition and infocommunication, creating a new environment for knowledge management. It is argued that the long evolutionary process of linear information processing constituting narrative-like mental structures based on cultural conceptualizations of any culturally coherent human community has already been challenged by parallel and connected network-based information processing making use of fragmented, encapsulated information chunks provided by a plethora of information sources. It is claimed that traditional learning conditions were constituted by a hierarchically determined distribution of knowledge and information and by a norm-based culture of behavior patterns. In opposition to the traditional learning conditions, the emerging patterns of digital cognition, digital information management and digital literacy are not only faster involving greater complexities, but they also facilitate coping with virtualization in general, leading to new cultural landscapes involving augmented reality. The new phenomenon is acknowledged in the paper by adopting the term *Connected Cognitive Entity Generation* (CCE Generation), by which I refer to the participants of the new information management practices induced by the very nature of the digitalized information environment. It is also claimed that the post-Generation Y is waiting in the wings, having finally broken out of the hierarchically structured, authoritative knowledge distribution due to the unprecedented advancement and development of the digital technologies. As opposed to Generation X or Generation Y, cognitive entities constituting the digital space as 'digital natives' will continually generate connected cognitive entities (CCEs) who will no longer be confined to generating a demographically conceived population. They will be social actors as interacting cognitive entities.

I will elaborate on the observation that cognitive entities in the digital environment are related to each other, not by commonly shared cultural narratives (as stated in the tenets of cognitive cultural anthropology to be depicted below in the following sections) but by random interest in networking, information sharing, and emergent cognition. One should also realize that this type of networked information exchange is not teleology-driven. What we witness under such conditions is an exponential growth of potential information sources due to the parallel design of connections.

After a brief survey of the different approaches to literacy, I will look at the contribution of education – formal and informal alike – and that of socialization – primary and secondary – to the development and shaping of literacy. Based on the conceptual framework I have adopted and the arguments I set out to elaborate on, I will put forward a weak hypothesis and a strong hypothesis.

The weak hypothesis assumes that there is no sharp dividing line between familiarity with disciplines in the natural sciences and with disciplines in the humanities and social sciences from the point of view of information management. I presume the presence of universal mental processes in perceiving, processing and managing information of any type (be it verbal, multi-modal, concrete, abstract, figurative, symbolic, analogue or digital, etc.).

The strong hypothesis claims that there are crucial and decisive differences in how the universal mental processes are put to use with the help of flexible adaptations to context-sensitive or context-dependent parameters of mental modeling. Exploiting the vast potentials and increased efficiency of individual and social cognition provided by digital information technologies will be a landmark in learning theories. However, the development of digital competences might involve a danger at the same time by the creation of a demarcation line or socialization gap between the advantaged and the disadvantaged. The use and misuse, or the exploitation and the lack of exploitation of a new world of information flow and information processing might create the greatest challenge for educators and politicians in finding the right ways and pace of social integration in the near future.

13.2 Cultural Conceptualizations and Cultural Narratives

13.2.1 Background assumptions

Knowledge management has been in the focus of interest for a few decades, however, consensus among representatives of cognitive anthropology, cognitive linguistics, artificial intelligence and cognitive infocommunications culture shaped by digital communication environments has been scarce. On one hand, this can be accounted for by the different disciplinary traditions pertaining to the respective fields. On the other hand – what the theoretical orientation of our research strongly suggests – it is to be acknowledged that the difference lies in the fundamental presumptions about social cognition within the realm of cognitive anthropology and those within the realm of digital culture. My objective in this paper is to point at the crucial difference in the assumptions. It is exactly the new assumptions regarding cognition and learning in the digital culture that seem to have led to a paradigm change as an inherent property of the digital era.

For the sake of convenience and transparency, I propose a comparison between two fascinating fields of research profoundly committed to knowledge management in the social space. On one hand, I will refer to current tenets of cognitive anthropology by discussing the concept of *cultural conceptualizations and situated cognition* manifest in socially situated discourse (Frank, 2008) and the concept of *distributed cultural conceptualizations and emergent cultural cognition* (Sharifian, 2003; Sharifian, 2008; Sharifian, 2011). On the other hand, I will survey the concept of *digital communication environments creating augmented virtual realities* (Abrash, 2014) and the concepts of *cognitive entity generation* (Baranyi & Komlósi, 2015; Baranyi, Csapó & Sallai, 2015).

13.2.2 Collectively constituted linear narratives with commonly shared interpretational patterns

We have witnessed a shift in cognitively oriented studies from the individual perspective on cognition to the social and cultural embeddedness of both linguistic cognition and social cognition, which has stimulated and widened the interest of researchers to a great extent. The idea of sociocultural situatedness revealed the mechanisms of cultural conceptualization and came to be seen as the foundation of socialization and interactive communicative behavior. An outstanding tradition in this respect has been established by the representatives of *Conceptual Metaphor Theory* (Lakoff & Johnson, 1980; Kövecses, 2005; Kövecses 2006).

The socio-cultural frame underlying the interactive communicative patterns of a community can be perceived as a mental framework providing a virtual fabric, i.e., a cohesive narrative to accommodate and determine culture-specific interpretation schemes for the members of cultural groups. Frank (2008) describes *socio-cultural situatedness* as a vital interaction of the individual mind with social and cultural structures (such as other agents, artifacts, conventions, etc. – in short, tangible and intangible cultural participants), thus facilitating social cognition involving learning, socialization, collective memory, mental maps of events, norms and patterns of behavior, etc. Although the nature of the interaction in the theory of socio-cultural situatedness is not spelled out in detail, one cannot fail to observe the quintessence of *situated cognition*: the socio-cognitive involvement of the individual mind is realised via interactions with other individual minds and with other social and cultural entities, where the community of individual minds collectively creates cohesive and supportive narratives in the course of these interactive processes.

One cannot fail to evoke at this point the foundational philosophical tradition established by Habermas's consensus-seeking communicative acts (Habermas, 1994). His starting claim is that discourse has a socially situated nature which determines and restricts communication as an interactive meaning-creating activity to be realized within the horizon of shared, unproblematic convictions, which automatically constitute consensus-generating interpretative patterns (Habermas, 1994:66). It entails – at the same time – that communicative actors are always moving within the horizon of their *lifeworlds* since they cannot step outside of it. Lifeworld is the invisible and indispensable background of everything we do and of everything we are.

Frank (2008) succinctly summarizes the outcome of the community of 'lifeworlds', which are the results of fairly homogeneous cultural conceptualizations: their manifestation is a culturally transmitted and linguistically organized stock of interpretative patterns. Lifeworld can be understood as a kind of non-thematic knowledge that is characterized by an unmediated certainty and a holistic constitution. It is composed of cultural patterns, legitimate social orders and personality structures, forming complex contexts of meaning.

Sharifian (2003) proposes a complex theory of cultural conceptualizations that function as collective representations to secure permanence in culture. This process, however, is dynamic and highly interactive within a cultural community as conceptualizations are represented in a distributed fashion across the minds of a cultural group. Cultural dynamics manifests itself in discourse the condition of which are bound to be permanently modified as the members negotiate and renegotiate these conceptualizations across generations. It is important to distinguish between the terms *narrative* and *discourse*. In Sharifin's theory of *distributed, emergent cultural cognition* (Sharifian, 2008; Sharifian, 2011), collective narratives secure permanence in culture, while discursive processes permit the renegotiation of cultural conceptualizations across generations. It is a novelty in the formulation that renegotiated and augmented cultural narratives reveal emergent properties in the flow of cultural interactions. Sharifian's dialectically integrative view sees cognition as a complex adaptive system that emerges from the interactions between the actors of a cultural group who negotiate and renegotiate their emergent cultural cognition across time and space. Although this theory offers an interesting perspective on augmented cultural narratives that overarch generations, it does not explain what exactly guarantees permanence in collective cultural representations.

It is to be acknowledged that collectively created cultural narratives have had their decisive function in shaping social cognition and social behavior across generations functioning in a linear mode of information processing. These narratives – as results of the complex adaptive cognitive systems of the members – have been successful in creating culturally controlled and augmented realities by allowing interaction and renegotiation of emergent cultural cognition. These cultural competences enabled, orientated and encouraged members of the cultural group to cope with hierarchically built, conceptually derivable (ideologically driven) meaning structures, to develop binding forces of cultural belonging and to seek permanent roles in given hierarchical structures.

The discussion that follows is meant to challenge Sharifian's attempt to support an integrative view of social cognition based on homogeneous cultural conceptualizations. The outcome of my analysis is to suggest that permanence in collective cultural representations cannot be secured under the circumstances of contemporary digital environments producing *a plethora of fragmented narratives* due to the alternative, parallel mode of information processing. Digital culture enables cognitive entities in the digital environment to cope with network-generated, associative meaning structures and to make sense of fragmented narratives and emergent contexts.

13.3 Fragmented Narratives in Digital Communication

13.3.1 Pragmatics: reflective contextualization and context-sensitive interpretations

There are many aspects of social interaction, however, the notion *contextualization* seems to be a central and unavoidable ingredient of pragmatic studies. Building contexts or observing and acknowledging contexts are results of creative mental activities. In some sense, we can assert that contexts are representations of states of affairs and situations. We can talk about the ontological foundation of representations or conceptualizations, and we can talk about their epistemological foundation as well. I proposed a systematic hierarchical relationship between foundational settings, ontological commitments and epistemic states in connection with contextualisation in (Komlósi, 2011; Komlósi, 2012):

1. Situations and faithful mappings of situations
2. Contextualized situations (selective mental representations of situations)
3. The linguistic context (texts and discourse depicting contextualized situations)
4. The pragmatic contexts (constructed contexts based on users' perspectives)
5. The context of social interaction and culture (social reality, knowledge of others)
6. The context of the self (figuring in individual and social cognitive situations)
7. Instantiated mental contexts (situated language use)
8. The context of the web experience (cognition in virtual reality)

I claimed that 1–2 manifest *situational contexts*, 3 represents a *linguistic context* and 4–8 realize *mental contexts*. Social interactive skills require and entail those higher level cognitive skills that facilitate adaptive, context-sensitive interpretations. Here, we have to observe that *social cognition* is a cover term for a vast number of different cognitive activities responsible for information processing and knowledge management with relative independence of the types of informational inputs. A *mental context* is constructed out of selected properties of mental contents, thus serving as background for evaluating the meaningfulness of particular arrangements in information structure. Linguistic structure in itself is not sufficient to determine contextual meaning. A mental context yields added value with which linguistic meaning is to be complemented.

13.3.2 Fragmented narratives

An array of contemporary research suggests that we are experiencing a new paradigm in perceiving, conceiving and managing information, especially with those people who have been socialized in the digital era and have internalized the very nature of digital culture. The members of the digital community function in connected networks created by a number of different types of cognitive entities. These people are cognitive entities themselves who feel at ease with digitalization and virtualisation (Abrash, 2014). In their environments, all sorts of smart devices are at their disposal such that they can cope with a plethora of information and can communicate effectively in accordance. We can assert today that these social actors function as cognitive entities. The notion *cognitive entity* (Baranyi & Komlósi, 2015; Baranyi, Csapó & Sallai, 2015) encompasses both human and non human agents, digital and mechanical entities that behave smartly and who/which are presumed to exist within a highly interlocked framework of a multi-channel information flow and exchange.

Cognitive entities in a digital environment are related to each other not by commonly shared cultural narratives (as stated in the tenets of cognitive cultural anthropology earlier), but by random and spontaneous interest in networking, information sharing and emergent cognition. We may say that this type of networked information exchange is not teleology-driven. What we witness under such conditions is an exponential growth of potential information sources due to the parallel design of connections.

To continue the metaphorical description of 'collectively constituted cultural narratives', in the case of the community of cognitive entities, we can talk about *chunks of information* that might be seen as *fragmented narratives*. Any piece of information can be related to any other information by creating suitable or agreeable *contexts for interpretation* (see the description of *mental context* above). Therefore, we can still envisage information chunks that might fit into narratives of some sort. However, the narratives created online are not based on negotiation, let alone renegotiation. They are results of non-deterministic interpretations accepted for current states of affairs. These information chunks might be ephemeral and seemingly arbitrary. Creating contexts for interpretation means an innovative challenge for the interactive agents. Fragmented narratives become flexible building blocks for novel contexts. As opposed to cultural narratives, which are based on cultural conceptualizations and are linear in design, fragmented narratives are non-linear, non-deterministic and distributed in a parallel design.

It is important to observe that the social perspective of cognition adopts the view that human cognitive abilities, especially human intelligence, are emergent properties. Such properties originate both from embodiment (i.e., mental attributes are inseparable from the physical, biological, neurophysiological bases of human existence) and from the interaction between humans, devices, artifacts, infrastructure and environment. Understanding the nature of co-evolving and emergent cognitive capabilities will legitimate the concept of *cognitive entity* to be developed further in the near future.

13.4 Emergent Properties in Learning and Knowledge Acquisition

This study is to show that the effect of digitization on the social network of connected cognitive entities holds many implications. Social cognition is to be understood as a permanent learning process, which requires higher-level cognitive skills to process and integrate emergent properties arising in the digital space through the interaction of connected cognitive entities. I want to emphasize the innovative drive which resides in the interactive informational frameworks, which consist of a multitude of connections of a cognitive personality to other cognitive entities, human and non-human alike.

13.5 Digital Literacy and The Digital Gap – Challenges for Traditional Socialization Patterns

As we have seen from the analysis in the previous section, networked digital communication environments create novel conditions for learning and knowledge management. The cognitively determined personalities in an emerging network of smart cognitive entities facilitate non-linear, multidirectional, horizontal communicative interaction among the social actors and substantially weaken the vertical, authoritative dependency lines in communication and social behavior.

It has become widely accepted that we talk about the emergence of a new generation of cognitive entities or generation CE in analogy with the generations X, Y and Z. The members of generation CE are different if not unique in the sense that both their primary and secondary socialisations take place in an environment that is inseparable from cognitive infocommunications in the digital space.

As a result of my investigation, I am claiming that situated discourse based on cultural conceptualizations functioning as a uniquely determining aspect of social communicative acts has lost its explanatory power due to the heterogeneous character of the information sources in the digital environment. No exclusively determining cultural conceptualizations exist anymore, which would function as the ontological basis for shared convictions.

However, I have a piece of good news to share, too. I can claim that connected cognitive entities do share a *common ground* for communication, which draws on the emergent properties of infocommunications, i.e., the thrill of being continually connected to and being engaged in a throbbing digital environment.

We need to realize that the long evolutionary process of linear information processing resulting in and constituting narrative mental structures based on cultural conceptualizations of any culturally coherent human community has already been challenged by parallel and connected network-based information processing making use of fragmented, encapsulated information chunks provided by a plethora of information sources.

Creating contexts for interpretation means an innovative challenge for the interactive agents. Fragmented narratives become flexible building blocks for novel contexts. Whereas cultural narratives based on cultural conceptualizations are linear in design, fragmented narratives are non-linear, non-deterministic, and are distributed in a parallel fashion by design.

All these promising aspects may unfold if the respective societies provide for supportive social environments and appropriate infrastructure for all. Here, we have to be on our guards and see that the widely experienced socio-economic gap should not be accompanied by a growing digital gap. In this paper too, I intend to call attention to the undesirable phenomenon of the *digital gap in education* closely connected to the phenomenon of marginalization in terms of socio-economic status.

The role and responsibility of the educational system has to grow to avoid the negative consequences of exclusion and marginalization of certain social groups in society.

13.6 Conclusions

A report on contemporary research on the nature and consequences of digital culture seems to be timely as the scientific communities want to know more about the unprecedented development and innovation in information and communication technologies, which have already exerted unforeseen impact on social cognition, information processing and human learning. Due to the appearance of smart systems of various sorts whose interactive elements include reasoning entities with both human and non-human properties, it is to be expected that new ways of learning and knowledge management will be required soon.

The paper attempts to identify the causes of a paradigm change in social cognition and infocommunications, which creates a new environment for knowledge management. It is argued that the long evolutionary process of linear information processing constituting narrative-like mental structures based on cultural conceptualizations of any culturally coherent human community has already been challenged by parallel and connected network-based information processing making use of fragmented, encapsulated information chunks provided by a plethora of information sources.

The paper offers a comparison between traditional learning conditions, which were constituted by a hierarchically determined distribution of knowledge and information and a norm-based culture of behavior patterns and the emerging patterns of digital cognition, digital information management and digital literacy. The study reports on the new phenomenon called *Connected Cognitive Entity Generation* (CCE Generation).

The main argumentative thread of our research is that we see digital culture as a novel environment for social cognition that inevitably undermines the permanence of cultural narratives. Cognitive entities seek connection to each other in a parallel, but non-deterministic fashion in order to exploit the immense potential in a plethora of information sources. The consequences for the nature and mechanisms for social cognition based on digital culture cannot be predicted in a reliable way as of today. A new perspective on and an unprecedented practice of information management is, however, a part of an irreversible process.

The paper identifies the characteristics of the post-Generation Y in relation to information processing and knowledge management in order to support the growing awareness of the digital age and the assumption that socialization and acculturation in the digital era rests on a significant paradigm change in the realm of social cognition. Digital culture is claimed to be acquired by the members of modern societies in a smooth and seamless way, the result of which is the permeability between human agents and non-human agents in the contexts of information processing and communicative interaction. The paper calls attention to the undesirable phenomenon of the 'digital gap in education and socio-economic status' which is a consequence of exclusion and marginalization of certain social groups in society.

References

Abrash, M. (2014) "What VR could, would, and certainly will be within two years," USA: Valve Corporation /http://blogs.valvesoftware. com/abrash/. Retrieved on 26.09.2015

Baranyi, P. and A. Csapo (2012): Definition and Synergies of Cognitive Infocommunications, Acta Polytechnica Hungarica Vol. 9/ 1, pp. 67-83

Baranyi, P. and L. I. Komlósi (2015): "Connected Cognitive Entity and the CE Generation: Emergent properties in the digital space," (forthcoming)

Baranyi, P., A. Csapo and G. Sallai (2015): Cognitive Infocommunications (CogInfoCom). Berlin: Springer International Publishing.

Frank, R, M. (2008): "Introduction: Sociocultural Situatedness," in: Body, Language and Mind. Sociocultural Situatedness, Vol. II, R. M. Frank, R. Dirven, T. Ziemke, E. Bernárdez, eds., Berlin/ New York: Mouton de Gruyter, pp. 1-18.

Habermas, J. (1994): "Actions, speech acts, linguistically mediated interactions and the lifeworld", in: Guttorm Floistad (ed.) Philosophical Problems Today, Vol. 1. Dordrecht/Boston: Kluwer, pp. 45-74.

Komlósi, L.I. (2011): Contextualization and Cognitive Synergism. The Interaction of Ontology and Epistemology in the Interpretation of Contexts. In: Hölker, K. und C. Marello (Hrsg): Dimensionen der Analyse von Texten und Diskursen/Dimensioni dell'analisi di testi e discorsi. Festschrift für János Sándor Petöfi zum achtzigsten Geburtstag/Festschrift per János Sándor Petöfi in occasione del suo ottantesimo compleanno. Berlin/London/Zürich/Wien: LIT Verlag, pp. 186-203.

Komlósi, L.I. (2012) Linguistic Context, Pragmatic Context, Mental Context: Meaning Construction and Interpretation via Contextualization. In: Dontcheva-Navratilova, Olga and Renata Povolná (eds): Discourse Interpretation: Approaches and Applications. Newcastle upon Tyne: Cambridge Scholars Publishing, pp. 19-38.

Kövecses, Z. (2005): Metaphor in Culture. Universality and Variation. Cambridge: Cambridge University Press.

Kövecses, Z. (2006): Language, Mind, and Culture. A Practical Introduction. Oxford: Oxford University Press.

Lakoff, G. and Johnson, M. (1980): Metaphors we Live by. Chicago: The University of Chicago Press.

Sharifian, F. (2003): "On cultural conceptualizations," in: Journal of Cognition and Culture, Vol. 3.3. Leiden: Koninklijke Brill NV, 2003, pp. 187-207.

Sharifian, F. (2008): "Distributed, emergent cultural cognition, conceptaulisation and language," in: Body, Language and Mind. Sociocultural Situatedness, Vol. II, R. M. Frank, R. Dirven, T. Ziemke, E. Bernárdez, eds., Berlin/New York: Mouton de Gruyter, pp. 241-268.

Sharifian, F. (2011): Cultural Conceptualization and Language. Amsterdam: John Benjamins.

Karla Lupşan

14 On the Use of Hypermediality in Teaching Culture in German as a Foreign Language Context

14.1 Defining Hypertext and Hypermedia

Since the development of the World Wide Web in 1989 by Sir Tim Bernes-Lee and the Belgian engineer Robert Cailliau at the CERN European Laboratory for Particle Physics, people's lives have taken an increasing digital turn; certain age groups being especially Internet-savvy, and almost constantly online. As a result, it has become a part of our everyday life to constantly take pictures or make films and simultaneously share them via digital media, whether locally or remotely. This increasing *digitalization of communication,* therefore, not only dissolves the limitations of local and remote communication, but also leads to interpersonal interactions, which are media-assisted on a regular basis, thus compelling teachers to bring the digital media in face-to-face classroom sessions.

The permanent availability and versatility of digital media (the Internet, CDs/DVDs, interactive kiosks, mobile devices), provide, on one hand, new directions for scientific development, and on the other hand, offer scholars a source of new research topics and approaches. For Pisani, Radtke and Wolters (2014: 226), digital media share three common features: hypermediality, interactivity, and multimediality. Alongside computer science, these characteristics have become keywords in the research in the fields of communication, translation, education, etc. From the perspective of translation studies, Sandrini (2011: 235) points out that 'this development should not be seen as a mere change in media, but [...] [must] be analysed holistically, with all its implications at the level of text production and organisation, multilingualism, technical requirements and many further areas of investigation'.

The newness of the field is also demonstrated by the fact that, regardless of discipline, scholarly discourse, whether linguistics- or pedagogy-oriented, always begins by explaining the origin of the concept of hypertext (e.g., Foltz, 1996; Gerdes, 1997; Iske, 2009; Rehm, 2006; Schröder, 2013; Stahl & Bromme, 2005; Schulmeister, 2013; Tergan, 2003; Winter, 1998). The author of the present study subscribes to those theoretical perspectives, which distinguish between *hypertext* and *hypermedia.*

> *Hypertexts* represent a specific form of texts, since they present parts of the text by means of specific connections (hyperlinks), in mostly non-linear form. In this way, a network-like structure is created between the individual parts of the hypertext. If various media (images, texts, animations etc.) are connected together through hyperlinks, we speak of *hypermedia*. (Horz, 2015: 124)

Consequently, *hypermedia* will be understood as the multimediality of the documents (images, film, animations, text, audio, etc.), while *hypertext* will be used to denote the network-like connections of this multimedia content (e.g., Horz, 2015: 124; Iske, 2009: 1; Schröder, 2013: 77–79; Tergan, 2003: 335). Moreover, in what follows, the concepts *hypertext* and *website* will be regarded as synonyms (e.g., Schröder, 2013: 76). A concrete example of the above-mentioned concepts is the World Wide Web, understood as a global database of multimedia content (hypermedia), which can be connected by the users into constantly changing hypertexts.

In recent years, the use of the Internet and of hypermediality in teaching has been increasingly debated by pedagogical and psychological theories, which frequently emphasise the similarity of semantic networks – i.e., the fact that items of information are stored in the brain in a networked manner – to the construction of hypertexts:

> It is often assumed that the network-like representation of hypertexts provides an advantage for learning, since cognitivist models of long-term memory also speak of a network-like construction of knowledge representation. (Horz, 2015: 124)

14.2 The Didactic Approach

The theories presented in the previous section imply that it may be useful to train networked thinking. Moreover, in my own teaching practice, I have consistently been confronted with the fact that students almost never establish links to other knowledge or apply acquired knowledge to new problems, or else that they have difficulty in doing so. In order to develop these abilities, one very useful approach is to use projects, particularly as this teaching method represents 'a complex form of learning, which involves many subskills [...] which can also be described as methodological competencies, and can therefore [be] learned and practiced' (Emer&Lenzen, 2002: 193; Lang, 2009: 576).

The following competencies can be acquired and trained:

- Professional competencies (connections and facts related to the subject being dealt with...)
- Social competencies (teamwork, personal responsibility, groupwork, conflict-solving strategies...)
- Methodological competencies (strategies for finding a topic, for group formation, for literature searches, etc.)
- Organisational competencies (distribution and coordination of tasks, management of resources, such as time, money, space, energy, the organisation of field trips...)
- Skills (practical skills, the creation of brochures, working with computers...).

(Bundesministerium für Bildung, Wissenschaft und Kultur, 2001: 43–44)

A concrete example of the aforementioned competencies is the interdisciplinary teaching project entitled *(Country of choice) on the internet*, which extends over several semesters, and therefore, includes three sub-projects:

Overall project: (Country of choice) on the internet
Sub-project 1: (Country of choice) on the internet: (Country of choice) in pictures
Sub-project 2: (Country of choice) on the internet: create a hypertext/a website
Sub-project 3: (Country of choice) on the internet / (beliebiges Land im Internet): hypertext / website translation

The overall project can focus on any country, with the first sub-project suitable for teaching cultural knowledge and the latter two for language-based knowledge, such as teaching of text comprehension, text production (writing acquisition), or translation. It should be emphasized here that the demands on the students are initially low, and then increase gradually. Each sub-project builds prior content knowledge and trains skills which are necessary for the next sub-project. For example, the last sub-project, in the field of translation, presupposes certain cultural knowledge, as well as media competences, namely knowledge on the structure and generation of hypertexts/websites, since

> [...] the web [becomes] significant for translation, both as a research field and as an area of application. A new type of source text emerges, which brings with it new challenges generated by the internet medium as a form of communication, and the fact that it is constituted as a digital text. (Sandrini, 2011: 236)

Finally, the already acquired cultural, hypertextual and translation-relevant knowledge must be culturally adapted to the expectations of an anonymous group of foreign language speakers, which is by all means a challenging task for inexperienced translators.

14.3 The Cultural Sub-project

The cultural sub-project is entitled *Germany on the Internet: Reading images of Germany* (*Deutschland im Internet: Deutschland-Bilder lesen* - DBL) and extends over one semester, i.e., 14 learning units. The project aims at the active acquisition firstly of targeted professional and social competencies, and secondly of organisational competencies.

14.3.1 Learning goals

In short, this sub-project aims at achieving the following main learning goals:

1. Acquisition of cultural knowledge about Germany
2. Developing of media competency
3. Developing cooperation strategies

On one hand, the students are required to actively develop their cultural knowledge, at the same time as their research competency, and on the other hand, they improve their media competency. They are encouraged to discuss the subject-related content as well as reflect critically upon the use of the media, i.e., evaluate the quality of the Internet content. This learning method is suitable for the successful attainment of these goals since – as the constructivists would put it – this is active self-directed knowledge acquisition, which would not be possible using teacher-centered transmission of knowledge. It is this autonomous behaviour that allows the creation of complex connections between concepts and the processing of new information into lasting knowledge.

Moreover, the project method has resulted in the development of social competencies, and especially collaborative learning, which 'targets a collective interactive construction of meaning, and implies a commitment to a common goal' (Schulz – Zander, 2005: 129).

A secondary goal is the development of organisational competencies, such as, for instance, those related to planning, the distribution and coordination of tasks, and time management.

14.3.2 Procedure

If we exclude the first teaching unit, which is dedicated to formulating the objectives, to creating and coordinating the small groups and to planning, and the last unit, which is devoted to project evaluation, the remaining 12 teaching units can be divided into three stages of three units each:

1st stage: *Selection*: guided thematic Internet research in small groups
2nd stage: *Organisation:* processing the data into a combination of text and image
3rd stage: *Integration:* interconnecting the separate hypermedial units into a network

Selection: In the first stage, the students undertake an Internet search of images, films and texts related to certain cultural content items. They are instructed to search mainly for images and films which convey as much cultural information as possible in a concise manner. The task is also based on the psychological theories of learning with texts and images which stem from research on memory or text comprehension. 'The memory-oriented or product-oriented approaches (e.g., Kulhavy, Lee & Caterino, 1985; Mayer, 2009; Paivio, 1986)' show that 'information from text and images is better remembered when it is stored *simultaneously* – that is, with a greater degree of spatiotemporal and thematic overlap' (e.g., Ostermeier & Eitel, 2014: 21). Product-oriented approaches (e.g., Eitel & Scheiter, 2014; Ostermeier & Eitel, 2014; Schnotz, 2001) have pointed out that images and texts 'complement each other in the construction of an effective mental model', since an image facilitates the semantic interpretation of a text 'while reading a text facilitates in turn the choice (selection) of relevant information in the corresponding image' (Ostermeier & Eitel, 2014: 23).

The outcome of this stage should be a well-organised archive of image, film, audio, and text documents centered around the relevant cultural topics. The collaborative research process builds information, research and media competencies, and fosters collaborative learning.

Organisation: After each group has collected material on the cultural content they have chosen, such as, for instance, landscape, historical overview, sights, way of life, and so on, in the second stage, they are asked to integrate the material into a combination of text and images. In addition to the media-oriented competencies, this task also develops reflective skills. Finally, the resulting multimedia documents are integrated and presented to the other students as PowerPoint presentations so that each small group shares their multimedia text. This also provides presentation skills practice.

Integration: The last stage consists in connecting the individual multimedia texts into a coherent hypermedia structure, which consolidates the already practiced competencies. The final product is then presented in the class and can also be submitted to a student conference.

The knowledge acquired by the selection, organisation and integration of the cultural material naturally provides the prior knowledge indispensable for the next two sub-projects: creating and translating websites with cultural content.

14.3.3 A teaching unit

In pedagogy, it is agreed that learners 'should only tackle tasks when they have acquired a basic understanding of the underlying [...] principles and their application' (Renkl, 2015: 15). In order to provide students with a clear understanding of the work they are performing, the method of parallel texts has been imported from the field of translation didactics. The concept of parallel text refers to authentic texts which appeared in similar communication situations in the target culture, which can be classed as the same type of texts and which deal with the same or a similar topic (e.g., Göpferich, 1998: 184; Kautz, 2000: 97–99; Sinner & Wieland, 2013: 107).

Students are already familiar with parallel texts from their translation courses so that, in the first teaching unit of the first stage (*Selection*), they are shown a so-called parallel text to the required end project. Thus, they are introduced to an example of hypermedia structure on a cultural topic created by the teacher, but one that deals with a country other than Germany. In this specific case, it is a hypermedia unit dealing with the Republic of Austria.

In this way, the students acquire a general idea of the most important topics in the study of culture. Otherwise, the self-directed research of the databases available online could lead to disorientation, to their being 'lost in hyperspace' (Conklin, 1987:

28). In order to prevent this disorientation and help students better evaluate the relevance of the content, the presentation of a parallel text as well as the permanent access to an already existing model are essential.

The presentation of parallel texts also facilitates collaborative learning, as the students can develop from the very beginning a common picture of the design of their end product. Otherwise, the existence of different design ideas may impede collective work.

In addition to this, with the help of parallel texts students also acquire an initial idea of the relations of dependence among different semiotic systems: text, image, sound and film, and may develop combinations of text and image or sound and image.

Although the method of parallel texts is not new, by the use of a parallel text in teaching culture, the present study proposes the transfer of an auxiliary method from the field of translation studies to another discipline. Thus, it represents a concrete example of how the main objective of 'networked thinking' can be achieved, by showing how connections can be established with the already acquired knowledge, and how this knowledge can be used in multiple situations, regardless of their nature.

The parallel text planned for one of the units, *Österreich im Internet. Österreich-Bilder lesen (Austria on the Internet. Austria in pictures)* discusses the main points regarding theAustrian culture, by means of interconnected data from different semiotic systems, i.e., the items of information on Austria are given, like in a hypertext, without aiming to be exhaustive, and in any order.

The introduction to the topic is made by means of 'a visual enumeration' of the best known clichés about Austria. The concept of Austria, or Alpine Republic, is the node for the next information item, namely *the landscape*. After the presentation of the most popular Austrian holiday destinations, the statement 'But over time other territories also belonged to Austria' connects to the topic of *Austria's History*. Maps, images or historical photos are chronologically presented to illustrate certain historical events and personalities, such as, for instance, the first Kingdom, the House of Babenberg, the Ostarrichi Document, the first record on Austria, or the Habsburgs. This chapter also includes sound and film documents on Emperor Franz Joseph I. Next, other historical events up to Austria's accession to the European Union are presented. As a member of the European Union, Austria enriches the European cultural heritage, a statement which makes the transition to the next cultural item, presenting the country's sights. The tourist attractions of individual federal states are presented in a networked manner and with occasional humorous comments. A joke regarding the animosity between Germany and Austria represents the node, which leads to the last chapter, *die Deutschen und die Österreicher (The Germans and the Austrians)*. A short film is played, *Alles gleich anders – warum Österreich nicht Deutschland ist (Otherwise all the same – why Austria is not Germany)*, broadcast by ORF and now available on YouTube.

The presentation can be found in the *Appendix* and the complete slide show in PowerPoint *Österreich im Internet: Österreich-Bilder lesen* can be accessed from the following link: https://drive.google.com/folderview?id=0BxZDTdykBXrNYjFRY05nRFVCY2c&usp=sharing

14.4 Conclusion

By the use of hypermediality in the project-oriented teaching of culture, important transdisciplinary competencies can be developed, such as professional competencies, media competencies, information and research competencies, social competencies, etc. The outcomes of the learning process and the acquisition of competencies are enhanced by the active, autonomous involvement of the students and by the fact that the teacher only rarely takes on the role of source of knowledge.

References

Bundesministerium für Bildung, Wissenschaft und Kultur. (2001). Grundastzerlass zum Projektunterricht. Tipps zur Umsetzung mit Erlasstext. Wien. Available at: https://www.bmbf.gv.at/schulen/unterricht/ba/pu_tipps_4905.pdf?4dzgm2 – accessed May 22, 2015.

Conklin, J. (1987). Hypertext – an introduction and a survey. IEEE Computer, 20 (9), 17–41.

Emer, W., & Lenzen, K. D. (2002). Projektunterricht gestalten – Schule verändern. Projektunterricht als Beitrag zur Schulentwicklung. Basiswissen Pädagogik. Baltmannsweiler: Schneider Verlag Hohengehren.

Foltz, H. (1996). Comprehension, coherence and strategies in hypertext und linear text. In: A. Dillon, J.J. Levonen, & J.F. Rouet et al. (Eds.), Hypertext and Cognition (pp. 109–136). Hillsdale: Erlbaum. Available at http://www-psych.nmsu.edu/~pfoltz/reprints/Ht-Cognition.html – accessed May 23, 2015.

Gerdes, H. (1997). Lernen mit Text und Hypertext. Berlin: Pabst.

Göpferich, S. (1998). Paralleltexte. In: H.G. Hönig, P. Kußmaul, & A.P. Schmitt et al. (Eds.), Handbuch Translation (pp. 184–186). Tübingen: Stauffenburg.

Horz, H. (2015). Medien. (2nd ed.). In: J. Möller, & E. Wild (Eds.), Pädagogische Psychologie (pp. 121–149). Heidelberg: Springer.

Iske, S. (2009). Hypertext, E-Learning und Web-Didaktik. In: H. Macha, & M. Witzke (Eds.), Neues Handbuch der Erziehungswissenschaft, vol. 3. Gütersloh: Schöningh. Available at: http://stefan.iske-online.de/uploads/pdf/2008_iske_handb.pdf – accessed May 22, 2015.

Kautz, U. (2000). Handbuch Didaktik des Übersetzens und Dolmetschens. München: Iudicium.

Kulhavy, R.W., Lee, J.B., & Caterino, L.C. (1985). Conjoint retention of maps and related discours. Contemporary Educational Psychology, 10, 28–37.

Lang, Ch. (2009). Projektunterricht – was ist das? Erziehung und Unterricht, 6, 570–579. Available at: http://oebv.at/sixcms/media.php/504/lang.pdf – accessed May 22, 2015.

Mayer, R. E. (2009). Multimedia learning (2nd ed.). Cambridge: Cambridge University Press.

Oestermeier, U., & Eitel, A. (2014). Lernen mit Text und Bild. e-teaching. Available at: https://www.e-teaching.org/etresources/media/pdf/langtext_2014_oestermeier-uwe_eitel-alexander_lernen-mit-text-und-bild.pdf – accessed May 23, 2015.

Paivio, A. (1986). Mental Representations. New York: Oxford University Press.

Pisani, P., Radtke, S. P., & Wolters, W. (2014). Handbuch visuelle Mediengestaltung. (7th ed.). Berlin: Cornelsen.

Rehm, G. (2006). Hypertextsorten – Definition – Struktur – Klassifikation. Available at: http://geb.uni-giessen.de/geb/volltexte/2006/2688/pdf/RehmGeorg-2006-01-23.pdf – accessed May 22, 2015.

Renkl, A. (2015). Wissenserwerb. (2nd ed.). In: J. Möller, &E. Wild (Eds.),Pädagogische Psychologie (pp. 3–24). Heidelberg: Springer.

Sandrini, P. (2011). Das Medium WWW als translationsdidaktische Herausforderung. In: I. Ohnheiser, W. Pöckl & P. Sandrini (Eds.), Translation–Sprachvariation–Mehrsprachigkeit. Festschrift für Lew Zybatow zum 60. Geburtstag (pp. 235–247). Frankfurt am Main: Peter Lang. Available at: http://www2.uibk.ac.at/downloads/trans/publik/twebdidak.pdf – accessed May 22, 2015.

Sinner C., & Wieland, K. (2013). Eine translationswissenschaftliche Sicht auf Sprachmittlung im Fremdsprachenunterricht. In: D. Reimann, & A. Rössler (Eds.), Sprachmittlung im Fremdsprachenunterricht (pp. 93–113). Tübingen: Narr.

Schröder, T. (2013). Marketingstrategien auf Unternehmenswebsites im internationalen Vergleich. Tübingen: Narr.

Schulmeister, R. (2013). Hypertext. Geschichte, Systeme, Strukturmerkmale und Werkzeuge. (2nd ed.). In: M. Ebner & S. Schön (Eds.),Lehrbuch für Lernen und Lehren mit Technologien (pp. 212–226). Available at: http://www.pedocs.de/volltexte/2013/8331/pdf/L3T_2013_Schulmeister_Hypertext.pdf – accessed May 23, 2015.

Schulz-Zander, R. (2005). Veränderung der Lernkultur mit digitalen Medien im Unterricht. In: H. Kleber (Ed.), Perspektiven der Medienpädagogik in Wissenschaft und Bildungspraxis (pp. 124–140). München: Kopaed. Available at: http://www.lmz-bw.de/fileadmin/user_upload/Medienbildung_MCO/fileadmin/bibliothek/schulz-zander_lernkultur/schulz-zander_lernkultur.pdf – accessed May 23, 2015.

Stahl, E., & Bromme, R. (2005). Das Schreiben von Hypertexten im Unterricht. Ein forschungsbasiertes Konzept. Unterrichtswissenschaft, 33 (3), 212–226. Available at: http://www.pedocs.de/volltexte/2013/5794/pdf/UntWiss_2005_3_Stahl_Bromme_Schreiben_von_Hypertexten.pdf – accessed May 23, 2015.

Tergan, S. O. (2003). Lernen und Wissensmanagement mit Hypermedien. Unterrichtswissenschaft, 31 (4), 334–358. Available at: http://www.pedocs.de/volltexte/2013/6782/pdf/UnterWiss_2003_4_Tergan_Lernen_Wissensmanagement.pdf – accessed May 23, 2015.

Winter, A. (1998). Arbeiten mit Hypertexten. Unterrichtswissenschaft, 26 (1), 32–50. Available at:http://www.pedocs.de/volltexte/2013/7763/pdf/UnterWiss_1998_1_Winter_Arbeiten_an_und_mit_Hypertexten.pdf – accessed May 23, 2015.

Appendix

Slide 1

Österreich im Internet: Österreich-Bilder lesen

Slide 2

Mozart, Sissi, Schönbrunn, Wienerschnitzel, Sachertorte das sind meist die stereotypischen Bilder, die einem zu Österreich zunächst in den Sinn kommen.

Slide 3

Die *Alpenrepublik* (Landkarte: Gesamtfläche von 83.858 km^2 und 8,9 Millionen Einwohner) im Herzen Europas hat aber zu jeder Jahreszeit und für die verschiedensten Geschmäcker und Interessen viel mehr anzubieten als nur verstaubte Klischees. Österreich ist ein Land, welches schon allein wegen seiner traumhaften Landschaft zu den begehrtesten Urlaubszielen in Europa zählt. Zwei Drittel des Landes liegen in den Alpen, die sich vom Bodensee bis zum Neusiedler See erstrecken. Nur ein Viertel ist Hügelland und Ebene.

Slide 4

Über 60% des Landes gehören den Ostalpen an, die in drei Großräume gegliedert sind:

1. Die Nördlichen Kalkalpen, Wiener Wald und Bregenzer Wald. Zu den größten Berggruppen gehören *Lechtaler Alpen und Allgäuer Alpen.*
2. Die Zentralalpen - dazu gehören *Ötztaler, Zillertaler Alpen, Hohe und Niedere Tauern.* In den Hohen Tauern erhebt sich der größte Berg des Landes, der GroOglockner (3797m).
3. Die Südlichen Kalkalpen mit der Gebirgskette *Karawanken* und andere

Slide 5

Im Westen bildet der Rhein die Grenze mit der Schweiz. Der längste Fluss Österreichs ist aber *die Donau*, die das Land vom Westen nach Osten auf einer Länge von 350 km durchfließt. Der schönste Teil der Donau ist die Wachau und dies ist ein etwa 30 km langer Donauabschnitt zwischen Melk und Krems.

Slide 6

Österreich verfügt auch über zwei markante Seelandschaften, eine im Salzkammergut: *Mond-, Atter- und Traunsee*, die zweite in Kärnten: *Ossiacher, Millstätter* und *Wörthersee*. Die größten Seen sind aber der *Bodensee* und der *Neusiedler See*.

Slide 7

Zu Österreich gehörten aber im Laufe der Zeit auch andere Gebiete, als das Gebiet, das heute den österreichischen Staat bildet. Den ersten Staat auf österreichischem Boden bildeten die *Kelten* im 2. Jh. v Chr., als sie das *Königreich Noricum* gründeten. Es dürfte das Gebiet von Salzburg, Oberösterreich, Teile von Niederösterreich, Kärnten und große Teile von der Steiermark umfasst haben. Dann dehnten die Römer ihr Reich bis an die Donau aus. Das Gebiet war bis im Jahre 800 immer wieder von anderen Völkern besiedelt. Dann bildete aber der fränkische Kaiser Karl der Große die Karolingische Grenzmark zwischen Enns und Wien.
Die Grenzmark wird aber im 10. Jh. zum zweiten Mal entstehen und von der Familie der *Babenberger* übernommen. In den 270 Jahren ihrer Herrschaft hatten sie auf friedlichem Weg aus einem kleinen Land ein ansehliches Reich gemacht. Sie dehnten sich nördlich der Donau aus und erweiterten die Mark nach Osten und Süden, erbten die Steiermark und kauften die Städte Linz und Wels. Die Babenberger schufen das Fundament für einen eigenen österreichischen Staat. 996 taucht zum ersten Mal in einer Urkunde der Name Österreich als *Ostarrichi* (Ost-Reich) auf.

Slide 8

Als die Babenberger ausstarben, fielen deren Länder an *die Habsburger*. Um die Mitte des 14. Jh. erwarben sie Kärnten, Tirol und Teile von Vorarlberg. In den folgenden Jahrhunderten vergrößerten sie ihre Hausmacht durch eine kluge Heiratspolitik. Als Karl der V. alle habsburgischen Länder vereinigte, wurde diese Dynastie zur mächtigsten Europas. Später wurde dieses Reich auf mehrere Linien der Familie aufgeteilt, die österreich–deutsche und die spanisch-niederländische Linie. 1700 starben die spanischen Habsburger aus und mit dem Tod von Kaiser Karl VI. beginnt die Herrschaft der bedeutendsten habsburgischen Herrscherin Kaiserin Maria Theresia. Maria Theresia leitete viele Reformen ein, die sich zum Teil bis heute auswirken. Von ihren 16 Kindern sind die bekanntesten, Marie Antoinette, die als Königin von Frankreich während der Französischen Revolution hingerichtet wurde und Joseph II, dessen Regierungsstil als Aufgeklärter Absolutismus bekannt wird. Der am längsten regierende Monarch Österreichs (68 Jahre) war Kaiser Franz I., der im 19. Jh. das Kaisertum Österreich gründete. Dies endete 1867 mit der Schaffung der Donaumonarchie oder der sogenannten K.u.K. Monarchie oder Monarchie Österreich-Ungarn. Die Monarchie war uneinheitlich, was die Kulturkreise, Nationalitäten, Sprachen und Religionen betraf, aber auch in den wirtschaftlichen und sozialen Strukturen. Der wachsende Nationalismus im Vielvölkerstaat verursachte schwere Spannungen und die Ermordung des österreichischen Thronfolgers Franz Ferdinand in Sarajewo bildete den Anlass für den Ausbruch des Ersten Weltkrieges.

Slide 9

Ich möchte Ihnen jetzt durch einige historische Aufnahmen die Möglichkeit geben den Kaiser Franz Joseph, auf der *Hochzeit des Thronfolgers Erzherzog Karl mit Zita* zu begleiten, aber auch seine Stimme zu hören. Der Kaiser hat 1903 seine Stimme auf Wunsch der Wiener Akademie der Wissenschaften im Phonogramm-Archiv verewigt und dieses historische Dokument wollen wir uns jetzt anhören.

Slide 10

Franz Joseph starb am 21.November 1916, während des Ersten Weltkrieges und mit ihm starb auch das Reich. Eine ganze Ära ging zu Ende. Zwei Jahre später am 12. November 1918 wurde in Wien die Republik ausgerufen. Der *Friedensvertrag von St. Germain* (1919) brachte den Verlust wichtiger Gebiete, indem die Nachfolgestaaten anerkannt wurden. Die neuentstandene Republik hieß Republik Österreich und hatte die gleiche Größe wie heute.

Slide 11

Da es Feindseligkeiten zwischen den großen politischen Parteien gab, wurde Österreich in kurzer Zeit eine faschistische Diktatur. Am 12. und 13. März 1938 wurde der Anschluss Österreichs an das Deutsche Reich offiziell verkündet und *die Schlagbäume an den Grenzen* wurden geöffnet. Zwei Tage später feierte *Hitler* den Anschluss Österreichs an das Deutsche Reich. Das Land trug von nun an den Namen Ostmark.

Slide 12

Es folgte der Zweite Weltkrieg, der bislang den größten Konflikt in der Menschheitsgeschichte darstellt. Die materiellen Schäden des Krieges will ich jetzt durch einige dokumentarische Fotos belegen. Gebäude, die heute zu den Symbolen Österreichs gehören, waren einst Ruine. So z.B. war der *Stephansdom* völlig ausgebrannt (wie sie sehen, suchen Kinder in den Trümmern nach verwertbaren Gegenständen), *das Parlament* war auch getroffen; beim ersten Luftangriff der US-amerikanischen Flieger auf die Stadt Salzburg am 16.10.1944 wurde der *Salzburger Dom* schwer beschädigt und auch das *Mozart-Haus* erlitt dann große Schäden. Erst 1945 erklärte sich Österreich wieder als selbstständiger Staat (die Zweite Republik), blieb jedoch bis 1955 von Truppen der vier Großmächte (Frankreich, Großbritannien, Sowjetunion und USA) besetzt. Am 15. Mai 1955 wurde in Wien der Staatsvertrag unterzeichnet und so erlangte Österreich seine volle Unabhängigkeit und Souveränität wieder.

Slide 13

Heute ist Österreich Mitglied der EU, zu der es am 1. Jänner 1995 als 15. Mitgliedstaat beigetreten ist. (*Korfu* - Am 24. Juni 1994 unterzeichnet der österreichische Kanzler Franz Vranitzky in Korfu die Beitrittsakte seines Landes zur Europäischen Union. Von links nach rechts: Sektionsleiter Ulrich Stacher, EU-Chefverhandler Manfred Scheich, Franz Vranitzky und Alois Mock, Bundesminister für Auswärtige Angelegenheiten).

Slide 14 und 15

Als Mitgliedstaat der EU bereichert Österreich das europäische kulturelle Erbe. Kultur pur erlebt man in den Großstädten Österreichs, wie z. B. in *Wien* Hauptstadt und Stadt an der schönen, blauen Donau, wie sie der weltberühmte Donauwalzer von Johann Strauß Sohn besingt. Dieser Walzer wird von den Wienern schon nach der ersten Aufführung als heimliche Hymne ihrer Heimatstadt empfunden. Allerdings besingt der Walzer nicht den Donaustrom, sondern „die Alte Donau" (1867), denn zur Zeit der Komposition war die Donau noch nicht reguliert. Die sogenannte Innere Stadt war früher von Verteidigungsmauern umgeben. Diese Mauern wurden 1857 abgerissen und durch einen breiten Boulevard, die Ringstraße ersetzt. Die Ringstraße wird auch noch die große Prunkstraße genannt. Zu den bedeutendsten Gebäuden zählen: *das Parlament, das Rathaus, die Staatsoper, das Burgtheater, die Universität.*

Slide 16

Im Südwesten der Innenstadt liegt die *Hofburg*, eine unregelmäßige breitgelagerte Gruppe von Gebäuden verschiedenartiger Baustile. Formen des Barocks und der Renaissance beherrschen ihr Äußeres. Die Burg war durch mehr als sechs Jahrhunderte Sitz der Herrscher Österreichs und durch zweieinhalb Jahrhunderte Sitz der deutschen Kaiser. Sehenswert sind hier: Das Ephesos Museum in der Neuen Hofburg, das Museum der Völkerkunde, die Österreichische Nationalbibliothek mit dem *Prunksaal* (Barockkunst), *die Schatzkammer* der Hofburg, da kann man die *Kaiserkrone*, Krönungsornate und Hochzeitszeichen, Schmuck und Erinnerungsstücke aus dem habsburgischen Besitz bewundern.

Slide 17

Die Ringstraße führt auch zu zwei wichtigen Museen, *dem Kunsthistorischen und Naturhistorischen Museum*. Die zwei Museen sind nördlich bzw. südlich des Maria–Theresia–Platzes gelegen, der vom Maria-Theresia-Denkmal beherrscht ist. Das Zentrum der Wiener Innenstadt bildet der Stephansplatz vor dem *Stephansdom*. Dieser ist der bedeutendste gotische Bau Österreichs und zugleich Wahrzeichen des Staates. Andere Sehenswürdigkeiten in Wien sind: *Schloss Belvedere*, Sommersitz für Prinz Eugen von Savoyen (Barockmuseum), *der Prater*, ein Vergnügungspark, wo man *das Riesenrad* sehen kann (Wahrzeichen der Stadt Wien),

Slide 18

und *Schloss und Park von Schönbrunn*, ehemaliges Lustschloss und der Lieblingsaufenthaltsort des Kaisers Franz Joseph. *Die Gloriette* im Park ist eine klassische Säulenhalle, als Siegesmal geschaffen zur Erinnerung an die Schlacht von Köln, in der Maria Theresias Truppen das preußische Heer Friedrich des Großen besiegten. Heute gehört Schönbrunn zu den UNESCO Welterbestätten.

Slide 19

Zum UNESCO-Welterbe gehört auch die Kulturlandschaft Wachau in NÖ. Am Eintritt der Donau in das Weinland der Wachau liegt die Stadt Melk. Sie wird überragt vom *Benediktinerstift Melk*, einem der bekanntesten und prächtigsten Klöster Österreichs. Es liegt auf einem steil zur Donau abfallenden Bergrücken und ist nur von Osten zugänglich. (Das Kloster kann im Rahmen von Führungen, etwa 1 St. besichtigt werden). Melk war die ursprüngliche Residenz der ersten Herrscher Österreichs, der Babenberger. Die Ursiedlung der Stadt Melk geht auf die Römerzeit zurück und ist das Medelike des Nibelungenliedes.

Slide 20

Etwa 5 km donauabwärts von Melk liegt am rechten Ufer *Schloss Schönbühel*, sehenswert ist hier die *Rosalienkapelle*. An der engsten Stelle der Wachau ist auf dem rechten Ufer die *Ruine der Burg Aggstein* zu sehen. Gegenüber der Ruine Aggstein liegt der Ort Willendorf, wo die berühmte „*Venus von Willendorf*" gefunden wurde (ein Kunstwerk, das um das Jahr 25 000 v. Chr. entstanden sein sollte und das sich heute im Naturhistorischen Museum befindet). Weiter flussabwärts sieht man am linken Donauufer die *Teufelsmauer*, eine eigentümliche Felsenrippe, dann folgt das Haupttor der inneren Wachau der Markt Spitz. 2 km flussabwärts von Spitz erreicht man die *Wehranlage St. Michael* und gleich danach den Weinbauort *Weißenkirchen*. Sehenswert sind hier die alte Wehrkirche und das Wachau-Museum.

Slide 21

Nachdem die Donau hinter Weißenkirchen einen Bogen nach rechts beschreibt, folgt *Dürenstein*, der meist besuchte Ort der Wachau auch „Perle der Wachau" genannt. Wenn die Donau sich in weitem Bogen wieder nach links wendet, taucht das Benediktinerstift *Göttweig* auf, das zu den eindrucksvollsten Klosterbauten Österreichs gehört. Am nördlichen Ausgang der Wachau liegt dann *Krems* an der Donau, die älteste Stadt Niederösterreichs.

Slide 22

Zum UNESCO-Welterbe gehören auch *Hallstatt*, zusammen mit dem *Dachstein* und dem Inneren Salzkammergut. Am Salzkammergut haben die Länder Salzburg, Oberösterreich und Steiermark teil. Es erstreckt sich von Salzburg im

Westen bis zum Dachstein im Süden und im Osten ist es vom Almtal begrenzt. Das Kulturwelterbe, das auch die höchste Touristenattraktion des Landes ist, bleibt aber die Landeshauptstadt und Mozartstadt Salzburg. Die Stadt liegt zu beiden Seiten der Salzach.

Slide 23

Die Stadt liegt zu beiden Seiten der Salzach.
Den Mittelpunkt der Altstadt auf dem linken Salzachufer bildet der *Residenzplatz* mit dem schönsten und größten Barockbrunnen, dem *Residenzbrunnen*. Die Südseite des Residenzplatzes beherrscht *der Dom*. Auf dem Domplatz wird seit 1920 während der Salzburger Festspiele „Jedermann" von Hugo von Hofmannsthal aufgeführt.

Slide 24

Nahe zum Fluss, in der Altstadt befindet sich Salzburgs berühmteste historische Gasse, die *Getreidegasse* mit dem *Mozarts Geburtshaus*.

Slide 25

Über der Altstadt, auf dem Festungsberg steht die malerische *Festung Hohensalzburg*. Man erreicht sie entweder zu Fuß oder mit der Standseilbahn. Das Innere der Festung kann besichtigt werden. Die Burg wurde im Jahre 1077 gebaut, erhielt aber ihre heutige Gestalt größtenteils um 1500. Im 17. Jh. wurde sie zur Festung ausgebaut.

Slide 26

Am rechten Salzachufer liegen die neueren Stadtteile. Das eigentliche Zentrum der Neustadt ist der *Mirabellplatz*. Links vom Platz steht das *Mirabell-Schloss* (heute der Amtssitz des Bürgermeisters und des Magistrats) und südlich vom Schloss liegt der *Mirabellgarten*, ein ausgezeichnetes Beispiel barocker Gartenkunst (Terrassen, Marmorstatuen, Springbrunnen).

Slide 27

Die Grazer – Altstadt, ebenfalls eine UNESCO-Welterbestätte wird vom Schloßberg überragt. Auf dem Schloßberg steht das Wahrzeichen der Stadt, der *Uhrturm*. Ursprünglich verfügte die Uhr des Turms lediglich über einen einzigen großen Stundenzeiger pro Zifferblatt, was zur besseren Ablesbarkeit aus größerer Entfernung beitrug. Um den später installierten Minutenzeiger vom Stundenzeiger unterscheiden zu können, musste er daher kleiner gestaltet werden. Deshalb sind die vergoldeten Zeiger auch heute noch in der Größe vertauscht. Im Zentrum der Altstadt ist der Hauptplatz mit dem *Erzherzog Johann Denkmal* und dem *Rathaus*. Sehenswert ist hier auch das *Luegg-Haus* mit der Stuckfassade. Ein Renaissance-Juwel ist das *Landhaus*, heute Sitz des steirischen Landtages.

Slide 28

Zwar sind sie keine UNESCO-Kulturstätten, aber auch Kärnten, Tirol und Vorarlberg zählen eigentlich zu den beliebtesten Urlaubszielen des Landes.
Mittelpunkt des Tiroler Tourismus sind die Alpen und die Hauptstadt Innsbruck. Die populärste Sehenswürdigkeit der Stadt ist das *Goldene Dachl* (vergoldetes Kupferschindel, spätgotisch). 13 km östlich von Innsbruck befindet sich die Marktgemeinde Wattens, der Stammsitz des Kristallunternehmens *Swarovski*. Der Gründer des Unternehmens, Daniel Swarovski, entwickelte einen mechanischen Schleifapparat für Kristallglas. Seine Schmucksteine wurden von einem internationalen Bijouteriemarkt als „Schmucksteine neuer Qualität" aufgenommen und wurden schnell weltweit bekannt. Zum hundertjährigen Unternehmensjubiläum wurde 1995 in Wattens als Touristenattraktion das Museum „Kristallwelten" eröffnet. Die Eröffnung und seine Umgestaltung 2005 wurden von dem österreichischen Künstler André Heller vollzogen. Mittlerweile entwickelte sich das Swarovski-Museum zu einer der größten Touristenattraktionen in Österreich und belegt mit durchschnittlich 700.000 Besuchern pro Jahr die Nummer zwei bei den Museen hinter Schloss Schönbrunn in Wien.

Slide 29

Der touristische Anziehungspunkt von Vorarlberg ist der *Bodensee*, an dessen Ostufer sich *Bregenz* erstreckt. Ein Kuriosum ist das *Kleinwalsertal*. Es gehört zu Vorarlberg, ist aber nur durch Deutschland erreichbar, da es von Bergen umschlossen ist und keine Straßenverbindung zum übrigen Vorarlberg hat. Früher hat man hier die deutsche Währung benutzt.
Burgenland beherbergt aber erneut ein UNESCO-Welterbe und zwar die Landschaft Fertö-Neusiedler See. Der *Neusiedler See*, der größte See Österreichs und der einzige Steppensee Mitteleuropas, hat keinen natürlichen Abfluss, ist aber durch einen Kanal mit einem Nebenfluss der Donau verbunden.
Burgenland wird manchmal „das jüngste Kind Österreichs" genannt, da es erst im Jahre 1921 ein Bundesland Österreichs wurde. Vorher war es ein Teil Ungarns und hieß Westungarn. Nach dem Ersten Weltkrieg verlangten die Deutsch-Westungarn den Anschluss an Österreich. Der Friedensvertrag ermöglichte dies, nur blieb die Hauptstadt Ödenburg bei Ungarn (heute Sopron, Ungarn). Zuerst sollte dieses Gebiet nach den Städten Pressburg, Wieselburg, Ödenburg und Eisenburg, das *V*ierburgenland heißen, aber alle Städte kamen bei der endgültigen Grenzziehung nicht zu Österreich.

Slide 30

Ein Burgenländer war *Joseph Hyden*, ein Komponist, der zusammen mit *Wolfgang Amadeus Mozart, Ludwig van Beethoven* und *Franz Schubert* mit dem Begriff Wiener Klassik identifiziert wird. Ein Witz über die deutschen

Nachbarn lautet: Österreich hat der Welt eingeredet, dass Hitler ein Deutsche und Beethoven Österreicher wären und den Deutschen ist das wurscht.

Slide 31

Sie sprechen - fast - die gleiche Sprache, teilen mit den Ösis eine turbulente Geschichte, sie können sich gegenseitig nicht ausstehen und kommen doch nicht einer ohne den anderen aus: die Deutschen und die Österreicher. Welches ist der Unterschied? Der Deutsche würde den Österreicher gerne verstehen wollen, kann es aber nicht. Der Österreicher versteht den Deutschen, will es aber nicht. Das ist einer der zahlreichen Witze, die das österreichisch-deutsche Ressentiment beschreibt.
Der folgende Film wurde vom *ORF* übertragen und versucht auf die Frage: „Warum Österreich nicht Deutschland ist?“ zu antworten.

Slide 32

Auswahlbibliographie

Baedeker – Redaktion (2000): *Österreich*. Wien: Karl Baedeker.
Bauer, Rolf (1994): *Österreich. Ein Jahrtausend Geschichte im Herzen Europas*. München: Wilhelm Heyne.
Bundespressedienst (1998): *Religionen in Österreich*. Wien.
Bundespressedienst (2000a): *Tatsachen und Zahlen*. Wien.
Bundespressedienst (2000b): *Musik in Österreich. Von den ältesten Spuren bis zur Gegenwart. Ein Überblick*. Wien.
Corti, Egon Caesar/ Sokol, Hans (1990): *Franz Joseph. Im Abendglanz einer Epoche*. Graz-Köln: Styria.

Internetquellen

Bildarchiv der Österreichischen Nationalbibliothek. http://www.bildarchivaustria.at/pages/search/quicksearch.aspx (abgerufen: Mai, 2015)
Bilder und Texte zu Urlaub in Österreich. https://views.austria.info/. (abgerufen: Mai, 2015)
Filmarchiv. http://filmarchiv.at/show_content2.php?s2id=346. (abgerufen: Mai, 2015)
Historische Bilder. http://austria-forum.org/af/Bilder_und_Videos/Historische_Bilder_IMAGNO/Franz_Joseph_I. (abgerufen: Mai, 2015)
Österreich-Portal. https://www.oesterreichinstitut.at/lernmaterialien/online-angebote/oesterreich-portal. (abgerufen: Mai, 2015)
Österreich. http://de.wikipedia.org/wiki/%C3%96sterreich. (abgerufen: Mai, 2015)

Phonogrammarchiv. http://www.phonogrammarchiv.at/wwwnew/. (abgerufen: Mai, 2015)

Swarovski. https://de.wikipedia.org/wiki/D._Swarovski. (abgerufen: Mai, 2015)

Filme:

Alles gleich anders. https://www.youtube.com/watch?v=WtD5T8Kaob0. (abgerufen: Mai, 2015)

Kaiser Franz Joseph I.- Tondokument. https://www.youtube.com/watch?v=kp-jlDsI8sQ. (abgerufen: Mai, 2015)

Kaiser Franz Joseph I. – Hochzeit Karl I. und Zita von Bourbon-Parma. https://www.youtube.com/watch?v=_p9_SW4zg-E. (abgerufen: Mai, 2015)

Iulia Para

15 Online Communication – Netspeak

The Internet as a Facilitator for New Ways of Communication and the Impact on our Language

15.1 Communication and Online Communication

According to Merriam Webster Dictionary, communication is 'the act or process of using words, sounds, signs, or behaviours to express or exchange information or to express your ideas, thoughts, feelings, etc., to someone else; a message that is given to someone: a letter, telephone call, etc.', while Oxford Dictionary sees communication as 'the activity or process of expressing ideas and feelings or of giving people information; methods of sending information, especially telephones, radio, computers, etc. or roads and railways'.

As the Internet has entered our lives, a new kind of communication has emerged – the *online communication*. The term *online communication* denotes communication, reading and writing by means of networked computers, which suggests that computer mediated communication helps computer users interact in real time, with all of them operating their computers, simultaneously.

Mark Warschauer (2001) mentions *asynchronous computer mediated communication* (whereby people communicate in a delayed fashion by computer, using programs, such as e-mail), and the reading and writing of online documents via www (World Wide Web). The www is a way of presenting information and any kind of materials in an electronic format, in an easy to access and use way. One can use www to communicate with other people, send messages, use e-mail, take part in conferences, and so on. It is the Internet that facilitates it and there is nothing worse than feeling left behind, when everyone is talking about something and you are 'in the dark'; nothing is more frustrating than being computer illiterate. If you can use the Internet, you are 'potentially connected with everything' (Crystal, 2001a:195). David Crystal mentions the name of 'the creator of the Web, computer scientist Tim Berners-Lee devised it in 1990 as a means of enabling high-energy physicists in different institutions to share information within their field' (Crystal, D. 2006:14), and, figuratively speaking, it *invaded* other domains of activity; nowadays it is designed for multimedia interaction among computer users all over the world. Consequently, new words connected to the web have been coined – *netizen* (Internet users) and *netspeak* (the language spoken by the netizens, Internet jargon) as an alternative to *Netlish, Weblish, Internet language, Cyberspeak, Electronic discourse, Electronic language, Interactive online discourse, Computer-Mediated Communication*, and various computer related locutions.

Although many people use www, most of them have no idea as to where this abbreviation (www) comes from. It is called so because all of the different pieces of information are linked together, and one can move from one piece of information to another by simply clicking the mouse button. The information we need can be anywhere in the world and this way, the information on our www page may link to other information on a page in Bucharest, London, New York, Tokyo and so on. 'The Internet is an electronic, global and interactive medium, and each of these properties has consequences for the kind of language found there' (Crystal, 2001a:26). 'The Internet has undoubtedly appeared almost from nowhere to take a rather important role in our lives', Hadžiahmetoviü (2007:2); the Internet is meant to present a wide range of information and is, therefore, used by individuals, educational institutions and companies. It can be used for anything that other information systems are used for: presenting facts and figures, academic research, advertising and launching new products, disseminating news and ideas on a variety of topics, talk about ourselves, our businesses, or our interests.

We shall not enter into details concerning how it works from the technical point of view; we shall focus on the purely linguistic dimensions of the written expressions – the use of *vocabulary*, *spelling*, *grammar*, and other properties/features of the electronic discourse (netspeak). 'Whatever variety of written language we may have encountered in the paper-based world, its *linguistic features* have their electronic equivalent on the web. It has extended the stylistic range of the language in interesting and innovative ways' (Moise, 2004:529). The e-mail, for example, is an opportunity for linguistic education, and overall, the Internet has developed and diversified written communication. Everyday communication – at work, in schools, and in private, is Internet mediated. Understanding and making progress in this new type of communication in terms of style, lexis and register help the Internet users become skilled communicators. The language of the web is no longer under a 'central control'; it does not consider national boundaries; 'people have more power to influence the language of the Web than in any other medium' (Crystal, 2001a:208).

The main platforms people use for online communication are: *Facebook* – a medium used for posting pictures, video-clips, make comments, or write articles; it is also used as an opportunity to announce important events. In some universities, it is also used as a teaching aid – for example, Professor Casimir C. Barczyk of Purdue University Calument, USA (Barczyk&Duncan, 2014) uses Facebook for teaching and communicating with his students.

There are also platforms such as *Twitter* – used by people to exchange or share ideas, and opinions, and whatever one wants to communicate to one's friends and to the world (chit-chat), *Tumblr* – a blogging platform similar to Twitter, where users could share photos, videos, and make comments, *Snapchat* – an application that makes possible for one to post photos and videos for a short period, *Instagram* – a photo sharing application that is used for photos to be shared with others, mainly their group of friends. One of the features shared by most of these online communication platforms is that users can also go for a private or public account, and can share photos or videos directly with their group – friends or acquaintances; in a way, it is a sort of a private message, in a different form.

With *online communication* being such an important part of our lives, people, especially scientists – experts in communication, have started to study this phenomenon and ask questions related to its impact on people's communication skills. We should be concerned regarding the consequences of social media in the long-run, on the young generation's critical thinking, on their writing and on personal and intellectual communication skills, especially in higher education, where the level of expressing oneself should be higher than average. Impact does not necessarily mean a negative impact, and using online communication may have either a positive or a negative effect; we shall focus on the positive impact only, and see the negative effect only as less and less face-to-face communication. Accordingly, blogging, texting and twitting via Internet is real-time writing and faster communication. Young people use netspeak because 'their writing is real writing, because their writing is their voice' (Rusul). In fact, social media's power of connecting people will enable the young generation's thoughts to impulse and promote change in our world, and their impact and attitude may generate remarkable ideas and thoughts. Therefore, as people, especially the young ones, are so much addicted to this kind of communication, sociologists often speak about 'homo smartphonius'; every time they have new ideas or opinions, they feel the need to share them with their friends or group of netizens.

Writing in *social media* is the writing of the young generation. According to Wikipedia, social media are computer-mediated tools (as the forementioned ones) that allow people to create, share, or exchange information, ideas and pictures/ videos in virtual communities and networks. It is defined as 'a group of Internet based applications that build on the ideological and technological foundations of Web 2.0 and that allow the creation and exchange of user-generated content'.

According to Barnbrook (1996), the linguistic future of the Internet – computer mediated communication 'may become an obsolete notion, as Internet information and communication functionality migrates to other tools, especially mobile phones, smart phones and tablets' Barnbrook (1996:214). According to the same author, technology developments may have three major impacts from the linguistic point of view: 'new modalities and their effects on the nature of language and speech community; new modalities bringing languages and speech communities into contact with one another; English – the common language of the Internet will influence the vocabulary of its users' Barnbrook (1996:214).

The question at this point is: will the Internet end the traditional spelling as it is known, and whether creativity and suppleness of the language will diminish due to globalization, which imposes sameness. In the history of mankind, any breakthrough, especially if connected to technology, was accompanied by fears and reluctance. David Crystal gives some examples in his Encyclopaedia of the English Language (2009:424), mentioning that 'in the 15th century, the arrival of printing was widely perceived by the Church as the invention of Satan' (Crystal, 2009:424). The priests feared 'that the dissemination of uncensored ideas would lead to a breakdown of social order and put innumerable souls at risk of damnation' (idem). Therefore, measures were taken rapidly for the limitation of its potential evil effects and consequences. Approximately 50 years after Gutenberg's first Bible (1455) in Frankfurt, censorship was extended to other books by Pope Alexander VI (in 1501). About four centuries later, identical worries were expressed, regarding the necessity of censorship and monitoring at the time of the invention of the telegraph, later the telephone and also about broadcasting technology. There were fears that the telegraph would affect family stability and would also support crime; the telephone could destabilize the stability of the society; broadcasting would manipulate people and be used for non-orthodox purposes. All these inventions generated fears, anxiety and also debates and linguistic differences of opinion. The debates were related to rules and norms to be obeyed, as well as correct pronunciation, on clarity and right utterances, and also regarding the use of local accent, or in some cases – dialects.

A similar debate is being continued nowadays, regarding the use of Internet and the language of online communication. Our aim is, therefore, to discuss the impact of Internet on language in general, and also on peculiar languages particularly. We shall try to identify to some extent, a few linguistic characteristics of the so-called *electronic evolution* and see whether the revolutionary way in which we use the language online is becoming different from our daily linguistic behaviour.

There must be made a clear-cut distinction between the paper-based communication (letter writing, telegrams) and the online communication (netspeak). The traditional paper-based communication relies on rules and conventions; the older generations employ their knowledge instinctively and very rarely break these rules. Some psychologists would argue that this trend of not using handwriting – in fact who is writing letters nowadays?, does not stimulate creativity, the capacity of memorising. People read less and less, and write even less. If they have to write something or fill in a document, it is a very stressful effort because they are not used to handwriting anymore, and they are used to only sign ready-made documents. About 17 years ago, calligraphy (handwriting) as a subject matter has disappeared from the Romanian schools. Decades ago, it was a school discipline for pupils in the primary schools; now it no longer exists. According to scholars, handwriting supports the cognitive circuits in the brain and when writing something by hand, we express ourselves in longer and more elaborate sentences. When we type, we only communicate the essential and there is no poetry, sensitivity, imaginary, originality, nor inspired and imaginary communication; everything is dry and sticks to facts like in a scientific communication. People no longer try to find the right word to express their ideas and feelings, what James Joyce called 'le mot just'. In the case of the Internet equivalent of traditional writing – e-mails or texts – there is no tradition. Some multinationals see handwriting as the reflection of a candidate's personality; therefore, they do not accept typed / computer processed applications and they always insist on handwritten document. These companies have a graphologist who will analyse the handwriting of each applicant and will examine each letter. S/he will eliminate those candidates who do not 'fit' for the job, and even if the candidate, theoretically, may seem the best for the job, s/he is not selected for the interview. The graphologist will identify those features which 'betray' the candidate and do not make him/her fit for that position.

There are some letters, such as those expressing condolences or congratulations for promotion or for being appointed in a higher position, which are also compulsory handwritten. If we want to show our consideration for family members, close friends, or people we appreciate, we do not send a 'ready-made' greeting card for Christmas, Easter, or other holidays, but we write one ourselves. The recipient will appreciate it if we took the trouble to write it and word it ourselves. However, the young generation very often disregards these conveniences and use the e-mail even in such cases. Most people have started using the e-mail or texting less than two decades ago; however, most of them disregard the factors and rules they should obey and consider, so as to avoid misinterpretation and misunderstandings of their texts.

15.2 Netspeak

Netspeak is a compound noun – *speak* involves both writing and talking; it also comprises listening or reading. Netspeak refers to speaking in ways commonly used to converse in text on the Internet. According to *macmillandictionary.com*, netspeak is 'the special language, abbreviations, and expressions used by people when communicating using the Internet'. Netspeak, or the Internet language is 'written speech' (Elmer–Dewitt 1994), while Davis and Brewer (1997:2), quoted by David Crystal (2006:27) say that 'electronic discourse is writing that very often reads as if it were being spoken – that is as if the sender were writing talking' (Crystal, 2006:27). Some of the newly invented words belonging to netspeak are already included in the *Oxford Dictionary of New Words* 1997 (e.g. *e-text, e-zine, e-cash, e-money*). Other words are not yet included in this dictionary still, but we can come across them in online communication: *e-lancers, e-management, e-government, e-books, e-voting, e-newsletter, e-cards, e-shop*. The number of these new words that will still be in use in the future is impossible to predict, as language changes all the time. Netspeak – the type of language used on the Internet, displays some characteristic features which make it different from the traditional correct language; these features can be noticed if we study pieces of writing in online communication and are due to their electronic, global, and interactive character. Netspeak is mainly used on social media websites, online messengers, sms and very often in informal short e-mails. It is a code consisting of simpler or modified spelling of words in order to avoid wasting time while typing a message (texting). Messages are usually written as if someone were really talking (as mentioned in the two previous definitions), and might, therefore, not have the correct grammar or punctuation that formal writing should have (e.g., Marry: hey, bff how you doin? // Jane: im ok bff. how u doin?; Mark: ok, c2 next week??? // Derek: ttfn lol // Mark: haha ttyl).

As for the reasons why these linguistic inventions are popular, a large amount of abbreviated terms and acronyms are used to speed up texting by shortening words, or commonly used expressions into a couple of characters. Some of these new words ended up in becoming a trend, especially among teenagers. Emoticons are another variant of netspeak that uses specific characters that translate into an emotion (simplified facial expression); they are widely spread and something used to express agreement, disagreement, feelings in general, etc.

The most commonly used abbreviations are: u = you, ur = your, u r = you are, LOL = laughing out loud, BFF = best friends forever, IM = instant message, PM = private message, btw = by the way, OMG = oh my god, xoxo = hugs and kisses.

The most commonly used abbreviations in an SMS are: & = and; 2 = two, to, too; 2DAY = today; 2MORROW = tomorrow; B = be; BF = boyfriend; B4 = before; BRO = brother; BT = but; C = see; D8 = date; GF = girlfriend; GR8 = great; L8 = late; PLS = please; U = you; UR = your; ASAP = as soon as possible; CUL = see you later; HRU = how are you; LOL = laughing out loud; MU = miss you; IC = I see; X = kiss.

Owing to its popularity among youngsters, we have tried to identify some of the grounds that may explain why texting and instant messaging are frequently used:

- Texting is fast; it is phone-to-phone instant communication and it is not connected to the Internet, hence does not make use of a server, like in the case of e-mails. Consequently, it is sent to the addressee almost instantly and the reply may come within seconds. If a longer and more complicated text is to be transmitted, texting is not to be used; in such a case, the e-mail should be sent or a phone call should be made.
- Instant messaging is very much similar to face-to-face dialogue, by means of modern technology; therefore one may often see young people typing on Twitter, Facebook Messenger, or WhatsApp. It is an instant conversation device and youngers can be frequently seen to message while standing in line, during classes, or even when socialising.

In general, instant messages are preferred for quick answers, as phone calls take longer and are time-consuming.

- Due to smart phones, texting and instant messaging are easier, e.g., when the Iphone was launched, its advantage over the other competitors was that it had the video function enabling the user to see the interlocutor.
- People answer texts far more than phone calls or e-mails (according to specialists, 8 trillion texts were sent in 2014, 95% of those were read within the very first 3 minutes, while only 14% of phone calls to businesses were answered, and only 12% of e-mails were ever opened at all).

According to Moise J., Para I, Istodor D., (2014), if a business partner wants to get in touch with another party and also wants make sure s/he gets a reply, the best solution is to text.

- Texting gives both users flexibility to answer on their own time, after considering the problem, when stopping if driving/in the middle of something.
- We need to ask for some information on the spot or we have something important to communicate and we want to avoid disturbing people around us. We may be in the wrong place (church, theatre, workplace, on a bus, on a train, in crowded places), still we need to contact someone. Using the phone would be a sign of poor education if we disturb people around us and talk about private matters in public. In some venues, such as schools, universities, restaurants, churches, theatre halls, concert halls, it is forbidden to use the phone. For example, in Italy, in classy restaurants people are not allowed to use the phone, and even from the technical point of view, in some of them, it is impossible to use the phone, due to technical reasons (e.g., Verdi Restaurant, All' Alba AbanoTerme, Italy).

Instead of texting, sometimes only emoticons are used; these symbols will covey our state of mind, such as sad, happy, sorry, etc.

This kind of fast communication, which is very popular nowadays, has developed recently, and it is characterised by some changes in the way people communicate, i.e., it has some specific features which we will speak about in the next sub-chapter.

15.3 Netspeak Features

Internet vocabulary is, maybe, the most important netspeak features; however, it does not contain any terminology belonging to IT. New words and elliptical sentences are being used and they are only used in Internet communication. A resource for enriching the netspeak lexis is computer hardware: freeze, lock, down, crash; some others have emerged for the Internet users themselves: netizens, neters, netheads, surfers, wannabees. Some of them are neologisms, but most of them are everyday words used by netizens with a different meaning than the one we already know.

According to Crystal, (2001b), 'a popular method of creating Internet neologisms is to combine separate words to make a new word - compound (e.g. cyber-surfers)' (Crystal, 2001b:6). Some elements appear repeatedly: bug (*bugfix, bugnet, bugtracker*), mouse (*mouse pad, mouse click, mouseover*), click (*click-and-buy, one-click, double-click, left click, right click*), web (*webcam, webmail, webmaster, Webster* (a person that acts like a dictionary), *webhead* (web addict)), ware (*groupware, shareware, freeware, firmware*), hot (*hotlist, hotlink, hotmail, hotJava, hotspot*), the symbol @ (abbreviation coming from *accounting — at the rate of* has sometimes a prefixal function — *@ home, @ command*), blends (parts of different words joined make a new word, such as: *cybercide* (the killing of someone in a virtual game), *infonet, hypernet, netleg, netiquette,* *netizen, netnews, netspeak, usenet,* the substitution of a compound by an equivalent sounding entry: *e-mail* (electronic mail), *e-commerce* (electronic commerce), *e-cruting* (electronic recruitment), *dot* (dot address, dotcom organizations). Even Bill Gates, the co-founder of Microsoft and maybe the most well-known IT specialist, has published in 1999 a book called 'Business @ the speed of thought' (Gates, 1999), where he makes use of the @ sign. He also uses the @ sign instead of the preposition *at,* suggesting in a way the type and contents of the book – business and the Internet.

Besides the above mentioned examples, there are also individual coinages which characterise many conversations, but it is impossible to know them all, as usually they are popular within a certain community of netizens or a group of close friends.

Change of grammatical category — conversion may also be encountered, usually from noun to verb: to mouse, to 404/303 (be unable to find the page).

Abbreviations also characterise netspeak, as acronyms are very common: BBS = bulletin board system, BCC = blind carbon copy, DNS = domain name system, FAQs = frequently asked questions, HTML = hypertext mark-up language, ISP = Internet server provider, Letter + number combinations are also common (GO2Net, W3C = World Wide Web Consortium, P3P = Platform For Privacy Preferences), chat groups (which mainly consist of people who know each other pretty well may have their own abbreviations, although it is possible to find them in e-mails and web pages too).

Some of the most commonly used abbreviations, identified by David Crystal are: afaik = as far as I know, afk = away from keyboard, asap = as soon as possible, a/s/l = age, sex, location, atw = at the weekend, awhfy = are we having fun yet?, bbfn = bye bye for now, bbl = be back later, bcnu = be seeing you, b4 = before, bg = big grin, brb = be right back, btw = by the way, cfc = call for comments, cfv = call for votes, cm = call me, cu/cya = see you, cul / cul8r = see you later, dk = don't know?, dyr = do you remember, eod = end of discussion, f? = friends?, f2f = face to face, fwiw = for what it's worth, fya = for your amusement, fyi = for your information, gal = get a life, gmta = great minds think alike' (Crystal, 2006:91).

The smart mobile phones may have small screens and, therefore, new abbreviations have been 'invented'; sometimes the acronyms are not limited to words only, but also to sentences, such: AYSOS = are you stupid or something?, CID = consider it done, CIO = check it out, GTG = got to go, WDYS = what did you say? Individual words may also be reduced to just some letters: PLS = please, THX/TX = thanks, WE = whatever, B4N = bye for now.

Another important characteristic of netspeak is also its typical graphology and in general, all orthographic features are affected; an example is the random use of capital letters (when something needs to be emphasised) and the use of small letters almost everywhere. The principle 'save a keystroke' functions everywhere, as whole sentences may be written avoiding capitals or the absence of punctuation. Small letters are generally used; the use of capitalization adds extra emphasis, of underlines the importance of whatever needs to be communicated. However, in some contexts capitals are a must, as they may be abbreviations of institutions, organizations and they may be somehow confusing, as for example: domain names in web addresses and many others (e.g., Google, Wikipedia, UVT, FEAA, etc).

BRB: be right back; FWIW: for what it's worth; HTH: hope that helps; IDK: I don't know; IIRC: if I recall correctly; IMO/IMHO/JMHO: In my opinion/In my humble opinion/ Just my humble opinion; KWYM/KWIM: know what you mean/know what I mean; OTOH: alternatively; TMI: too much information; TIA: thanks in advance; TTYL: talk to you later; WTG: way to go; IRL: in real life; PM: private message; OMG: oh my god; PITA: pain in the ass; AWOL: away without leave; KISS: keep it simple, stupid; MIA: missing in action; DNR: do not resuscitate; OOP: out of power; RPG: role-playing game; WPA: well-played all; DIY: do it yourself; ARC: advanced reader copy; HEA: happily ever after; WIP: work in progress; DNF: did not finish; DQ: disqualified; FTW: for the win.

Netspeak also has a distinctive spelling practice. Since the Americans are the best in computers, U.S. spelling is more frequently used than British English, sometimes for reasons of economy (American words are one character shorter than the British ones – mediaeval/medieval; colour/color; labour/labor; behaviour/behavior). Innovations in spelling have also been noticed; non-standard spellings (e.g., tunez, filez, downloadz). Non-standard spellings are also used in informal conversations; spelling errors in messages or e-mails are not seen as a sign of poor or lack of education; many non-standard spellings imitate pronunciation (kay, sokay = OK; cheeze = cheese; iz = is; dat = that, tha = the, bai = bye). Several recent spellings introduced by youngsters could also be mentioned: *kool* for *cool*, or the use of small zero, or of the percentage sign (0 / %) for small *o*; the alternative use of *eh* for *y* (bunneh = bunny; funneh = funny; happeh = happy).

Another important characteristic of netspeak is open or no punctuation at all, being justified in a way as netspeak is a speech itself in a written form. Still, the use of punctuation depends on people, some are very careful in writing a message or an e-mail, while others do not use punctuation at all. Symbols ($ for s, £ for l) and emoticons are widely used (https://messenger.yahoo.com/features/emoticons/). Asterisks are also used to underline or emphasise the importance of something (It is a *very* *important* *issue*), while normally the sentence would be: It is a very important issue. The use of asterisks calls attention to that specific issue.

As it has already been mentioned, the most typical features of netspeak may be found in its vocabulary and graphology, where it is easy to introduce innovation. Grammar innovations are less frequent, and if they occur, they are restricted to a particular group of users. A typical example is verb reduplication, which means that a verb is written twice, in order to express great pleasure, astonishment, or exasperated reactions; sometimes, it is an indication that the conversation is ended (e.g., love it! love it!, go! go!, see you see you). A typical example of this characteristic is using capital P, placed after the word, thus turning the sentence into a yes/no question (e.g., cinema P = 'are you going to the cinema?'); however, this kind of language is distinctive of a particular group of users and cannot be extended to the Internet users in general.

:)	happy	>:)	devil	:-$	don't tell anyone
:(	sad	:((	crying	<:-P	party
;)	winking	:))	laughing	(:|	yawn
:D	big grin	:|	straight face	:-?	thinking
;;)	batting eyelashes	/:)	raised eyebrows	#-o	d'oh
>:D<	big hug	=))	rolling on the floor	=D>	applause
:-/	confused	O:-)	angel	:-SS	nail biting
:x	love struck	:-B	nerd	@-)	hypnotized
:">	blushing	=;	talk to the hand	:^o	liar
:-*	kiss	:-c	call me	:-<	sigh
=((	broken heart	:)]	on the phone	>:P	phbbbbt
:-O	surprise	~X(	at wits' end	X_X	I don't want to see
X(	angry	:-h	wave	:!!	hurry up!
:>	smug	:-t	time out	\m/	rock on!
B-)	cool	I-)	sleepy	:-q	thumbs down
:-S	worried	8-|	rolling eyes	:-bd	thumbs up
#:-S	whew!	:-&	sick	^#(^	it wasn't me

Figure 15.1: Emoticons.

The features of netspeak identified in the preceding section may not influence our language, but they can provide a firm starting point for taking into consideration, netspeak as an authentic language variety, as many scholars argue in its favour.

15.4 Conclusions

In this paper, we have tried to analyse the new and very popular form of communication, i.e., online communication and identify some of the main features of netspeak. Netspeak is a new kind of language, evidently related to English, and used as an instrument of communication; it has its features, which we have identified from several Internet sites, by speaking with our students, and of course, reading the literature related to netspeak, especially David Crystal's articles and writings. We have also mentioned the platforms people, especially teenagers, use for online communication, which are so often used that people have become addicted to smart phones and tablets. This electronic revolution has brought about a lot of changes; however, we do not consider that they will affect the language and the way we speak; most innovations in netspeak are used for the sake of texting faster, or 'saving a stroke'.

We have also noticed some modest attempts to use abbreviations and emoticons in Romanian SMSes and online communication. Our first contact with network was the name of the English band U2; then labels on presents to be given for Christmas, Valentine's Day, Mother's Day, Easter; and those labels have the short text *4U = for you*. Some firms have also started to use abbreviations (ETA 2 U – the name of an IT firm from Timisoara, Romania).

In general, netspeak is used by the young generation, especially in 'virtual socialising', as they are addicted to the new technologies. Even if netspeak may seem unintelligible for most of us, as it is far from what we consider standard writing, still this kind of *language* does exist and is used on a large scale, especially by the young generation.

Acknowledgement: This work was co-financed by the European Social Fund through Sectoral Operational Programme Human Resources Development 2007–2013, project number POSDRU/159/1.5/S/140863, Competitive Researchers in Europe in the Field of Humanities and Socio-Economic Sciences. A Multi-regional Research Network, sau Această lucrare a fost cofinanţată din Fondul Social European prin Programul Operaţional Sectorial pentru Dezvoltarea Resurselor Umane 2007–2013, Cod Contract: POSDRU/159/1.5/S/140863, Cercetători competitive pe plan European în domeniul ştiinţelor umaniste şi socio-economice. Reţea de cercetare multiregională (CCPE)

References

Barczyk, C. C. &Duncan, D. G. (2014). The 2014 WEI International Academic Conference Proceedings Vienna, Austria, Facebook in Higher Education: Contributions to Sense of Learning and Sense of Connectedness.

Barnbrook, G. (1996). Language and computers. Edinburg: Edinburg University Press.
Crystal, D. (2001a). Language of the Internet. Cambridge: Cambridge University Press.
Crystal,D. (2006). Language and the Internet. Cambridge: Cambridge University Press.
Crystal, D. (2009). The Cambridge Encyclopedia of the English Language. Second Edition. Cambridge: Cambridge University Press.
Davis, B. H.&Brewer, J. P. (1997). Electronic discourse: linguistic individuals in virtual space. Albany, NY: State University of New York Press.
Elmer-Dewitt, P. (1994). Bards of the Internet. Time, 4 July, pp.66-67.
Joyce, J. (2010). Ulysses. Wordsworth Editions Ltd.
Moise, J. (2004). The language of the Web, in Economia şi Managementul Transformării, al II-lea Simpozioninternaţional, Timişoara.
Moise, J., Para, I., Istodor, D. (2014). An Introduction to Communication in English, Timisoara, Ed. Mirton.
Oxford Dictionary of New Words.(1997). Oxford University Press.
Sanel Hadžiahmetoviü-Jurida, Jezikoslovlje 8.2 (2007): 193-210 ŋ 193 UDC 811.111'276.6:004.7 Original scientific paper Received on 26.10.2007. Accepted for publication 28.11. 2007.
Warschauer, M. (2001). Online communication. In R. Carter & D. Nunan (Eds), The Cambridge guide to teaching English to speakers of other languages (pp. 207-212). Cambridge: Cambridge University Press.
Webster Encyclopaedic Unabridged Dictionary of the English Language. (1996). New York: Merriam Webster Gramercy Books.

Webography:

Crystal, D. (2001b). Twenty-first century English (keynote speech to IATEFL Annual Conference, March 2001).www.davidcrystal.com/?fileid=-4108, seen 05/08/2015
Emoticons: https://messenger.yahoo.com/features/emoticons/, seen 10/08/2015
Rusul Alrubail. (2015). http://www.edutopia.org/discussion/social-media-students-communication-skills#comment-204736, posted 8/4/2015 8:07 pm, seen 15/08/2015

Alexandra Jic

16 Young English Learners in the Digital Age

16.1 Introduction

As Michael Fullan (2013: 23) well points out, students are increasingly bored in school and the only thing worse than being bored is 'having to teach the bored'. Therefore, he underlines the need to develop innovative solutions for the increasingly challenging educational landscape. According to him three, teaching directions could be the answer to reforming teaching in order to meet the needs of the 21st century learners: 'deep learning goals, new pedagogies and *technology*' (Fullan, 2013: 23). It is precisely this last element that represents the focus of the present paper in an attempt to address the importance as well as the teachers' main challenges of employing modern technologies to support English language learning of young pupils because 'Teachers are central to what happens in the classroom, because they have the knowledge and skills to find creative ways to support learners' language development' (Motteram, 2013: 178).

As most teachers would probably agree great transformations in the educational landscape are about to occur because society is changing at a rapid pace, and this is mainly caused by the advent and increasingly wide-spread use of technology in our everyday lives. Consequently, as Pim (2013: 18) emphasises 'learners are being exposed to a range of technologies from a *very early age* in the home and *by the time they reach nursery age many have developed at least some of the digital skills* that enable them to participate in technology-driven activities as soon as they start school'. Never before has learners' childhood been so intertwined with the use of technology. Adults are often startled at the children's capacity and adaptability to use the forever advancing technology. Hence, children's way of understanding and interacting with the world around them is directly influenced or sometimes even 'filtered' by technology.

Though many people frown upon this tendency, I firmly believe that teachers cannot turn a blind eye to the immense effect that technology has and will continue to have on our students, no matter we like it or not. Consequently, as many researchers (Wong L., 2013; Pim, 2013; Fullan 2007, 2013) have also pointed out, 21st century teaching should incorporate technology in the classroom and technology-based or technology-aided classes should represent pedagogical tools in order to enhance learners' motivation.

Within this context, the Romanian Ministry of Education is trying to keep up with the latest trends by trying to connect itself to the new educational methods of improving the educational process and adapting to the needs of the 21st century learners by bringing technology closer to school, informs *Mediafax*, one of the leading press agencies in Romania. (*http://www.mediafax.ro/social/pricopie-vrea-tablete-manuale-digitale-si-catalog-electronic-in-toate-scolile-13440664)*

Among the steps that have been made towards this end, the introduction of digital books for the 1st and 2nd grades starting from 2014 has probably been the most important as well as the most controversial one. Remus Pricopie – the former Romanian Minister of Education, who was in service until December 2014 – argues that Romania is one of the first countries that are reforming and adjusting their educational systems to the worldwide tendency of bringing technology into the classroom. According to him, this tendency is confirmed by the new The Horizon Report Europe, a publication of European Commission, which emphasizes the need for improvement of digital skills and of access to digital resources in order to enhance teaching and calls for urgent actions in order to promote innovation within the classroom. (http://www.edu.ro/index.php/pressrel/22251?theme=print)

However, as it has been widely noticed, in an attempt to reform the educational system, Romania has been undertaking a great variety of changes over the past 25 years, many of which have been heavily criticized as being chaotic and disorganised. The introduction of digital books was no exception and it has brought about as much praise as criticism. Although digital books were initially meant to be a great innovation for the Romanian educational system, there are voices who argue that they were introduced with insufficient consideration and prior planning. Therefore, the introduction of digital books presented teachers with both great benefits and multiple challenges because as Fullan (2007: 8) also emphasizes,

> "neglect of phenomenology of change - that is, how people actually experience change as distinct from how it might have been intended – is at the heart of the spectacular lack of success of most social reforms. It is also necessary to build and understand the big picture, because educational change after all is a sociopolitical process [...] in the process of examining the individual and collective settings, it is necessary to contend with both the "what" of change and the *"how"* of change".

The present paper, thus, focusses on teachers' perceptions of the way in which digital books were introduced, emphasizing the problems that teachers have to deal with in order to be able to use them in the classroom and the causes that might lie behind them because just as Wong outlines, 'Teachers are not only expected to keep up with new technologies but also to integrate IT into their curricula and classroom practices to equip students with the skills of the information age' (Wong L., 2013: 248). Moreover, the present paper represents an attempt to report on the needs of children learning a foreign language and on children's attitudes towards the English classes that incorporate digital books in the teaching and learning process.

16.2 Digital Books

The newly introduced books for the 1st and 2nd graders have both a digital version and a print one. Besides the images and the short written texts that are to be found in the print version, a digitized version integrates audio-visual media (an interactive book employing colourful visuals, animations, videos, short stories and songs). It is to be mentioned here that the digital version is not supposed to entirely replace the print version, but to complement it, the two being meant to function in parallel.

The purpose of digital books is obviously twofold: to support language learning both in the classroom and beyond it. They are expected to provide teachers with lively and interactive teaching materials meant to make the English class a fun and joyful experience.

However, in order for this to be possible, several conditions must be met: on one hand, in order for the teacher to be able to use the e-books in class, the classroom needs to have at least the basic educational technology, i.e., it should be equipped with a computer/laptop and ideally a projector and a screen to enlarge the image so that all the children may be able to watch the short videos, and thus, participate in the lesson. Moreover, the teachers should have at least basic knowledge of how to use this type of technology. On the other hand, in order for children to take advantage of digital books beyond classroom, they also need to have access to a computer at home.

16.3 Study

But what is the impact of using digital books within the Romanian educational system? In an attempt to identify the benefits as well as the challenges that digital books have brought about in the short period of time since their introduction at the end of the autumn of 2014, a both qualitative and quantitative survey was conducted at the end of January 2015, in a number of Romanian public schools. The findings are as much as surprising as revealing and call for immediate action in order to suit their initial purpose, i.e., supporting learning enhancement and student motivation.

16.3.1 Methodology and sample profile

In order to meet the above stated objectives, a paper-based survey was designed and administered to a number of 14 first- and second -grade English teachers working in the public sector. All responses were anonymous and the survey took no more than 15 minutes to complete. However, it is to be mentioned that the present study is limited to the Fairyland series as the Romanian Ministry of Education provided the English teachers with three options to choose from at the beginning of the first semester: *Magic English, Comunicare in limba engleza* (*Communication in English* – my translation) and *Fairyland*.

Teachers' profile:
- Gender: 93% females; 7% males
- Age: 35% aged 25–35; 50% aged 35–45, 15 % above 45
- Years of experience: 14%: 1–3 years; 22%: 4–10 years, 64% more than 10 years

Pupils' profile:
As far as the pupil sample is concerned, an informal interview was conducted over a number of 22 first-grade learners of English and 16 second-grade learners of English regarding their opinions on the use of digital books.

16.3.2 Questions for teachers and analysis

1. How would you rate your overall skill in using educational technology?
- Beginner
- Basic
- Advanced
- Proficient

Results

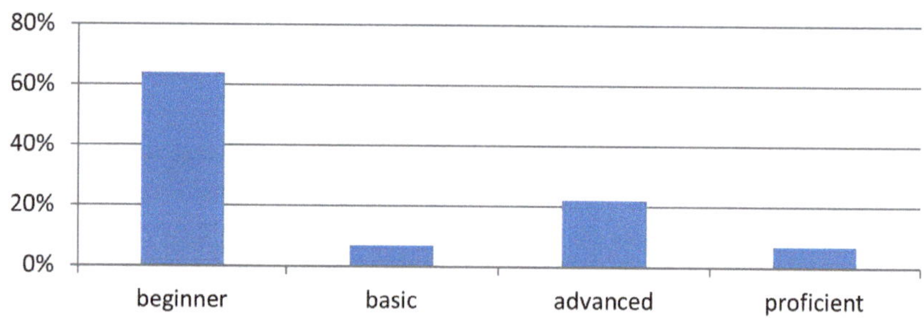

Figure 16.1: Overall skill using educational technology.

It is notable that an overwhelming percentage of teachers seem to be lacking the skills to use educational technology: 64% rated themselves as beginners, 7% basic, while only under 30 % (22% advanced and 7% proficient) seem to be equipped with the skills that allow them to use digital books in the classroom.

2. When was the last time you participated in a formal training organised by your school board/board of education in order to improve your information technology skills?
- Less than 1 year ago
- 1–3 years ago
- More than 3 years ago
- Never

Results:

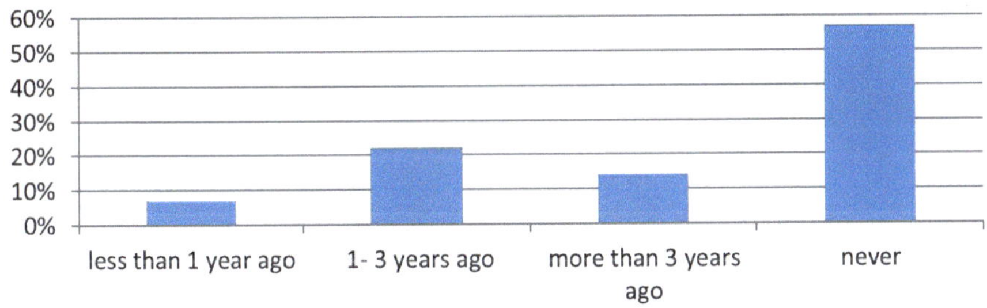

Figure 16.2: Last time I attended a training in IT.

It is noteworthy that almost 60% answered that they had never participated in an IT training. Unfortunately, only 7% of the teachers declared to have attended a training in IT skills within the last year. Analysing this in relation to the previous question, one could speculate that the teachers' poor command of educational technologies may be a direct cause of lack of training

3. Is your school equipped with one of the following educational technologies? Mark all that apply.
- computer/laptop
- projector
- smart board
- CD /DVD player
- none

Results:

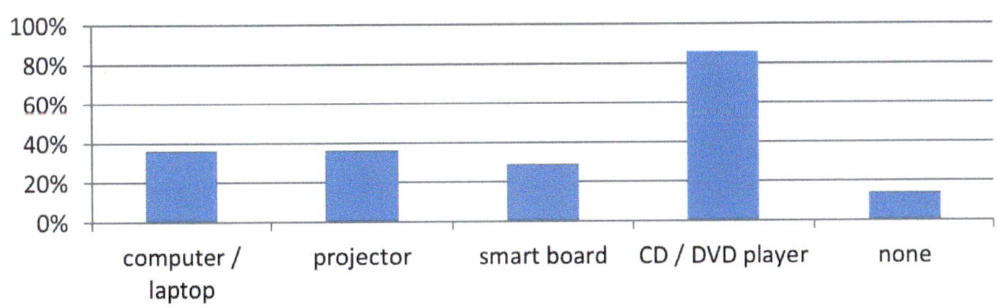

Figure 16.3: Educational technologies available.

Fortunately, it results that most schools are equipped with a CD/DVD player. However, having a CD/DVD player at one's disposal is unhelpful when dealing with digital books. In addition, it results that only about 36% of the respondents seem to have the necessary technology (a computer/laptop and a projector) to use digital books in the classroom. Regrettably, 14% of the respondents admitted not having any educational technology at their disposal.

4. How often do you integrate educational technology into your classroom instruction?
- Daily
- A few times per week
- Rarely. Please give reasons...
- Never. Please give reasons...

Results:

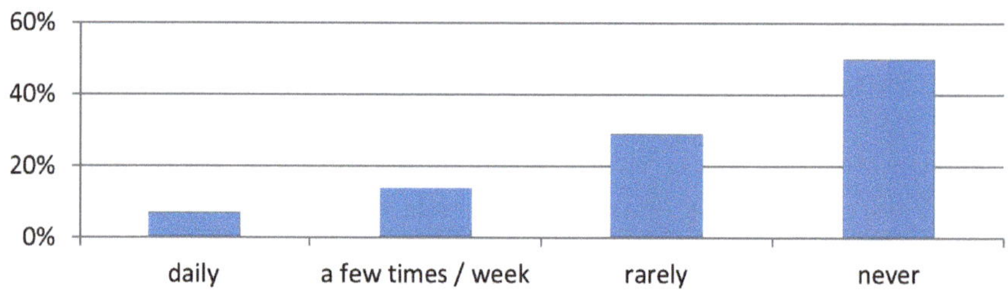

Figure 16.4: How often do teachers integrate educational technologies?

Unfortunately, only 21% of the teachers seem to integrate educational technologies on a regular basis (7% daily, 14% a few times per week), while 79% either do it rarely or never (29% rarely and 50% never). Three possible reasons were given by those in these two last categories: no or less availability of computers or laptops (they have to share them with other colleagues) and lack of knowledge on how to use them. So, it is not that they do not want to use them, but they either have no resources or no knowledge.

5. How often do you use the newly introduced digital books?
- Daily
- A few times per week
- Rarely
- Never

Results:

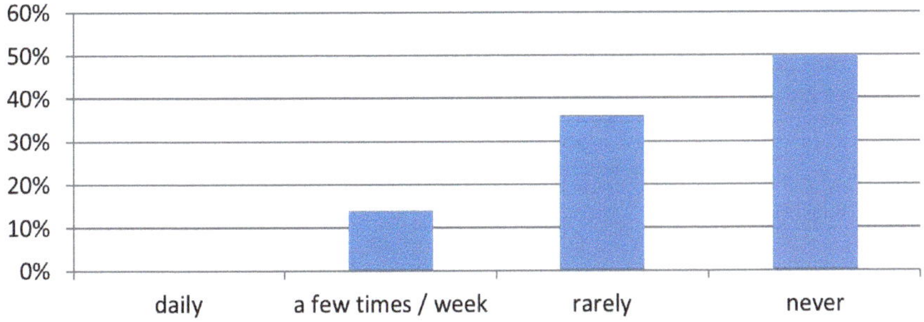

Figure 16.5: Use of digital books.

It results that most teachers either use digital books rarely (36%) or never (50%). Only 14% seem to use them a few times per week. What is particularly interesting when analyzing the answers to this question in relation to the answers to the previous one is the fact that none of those who do integrate educational technology on a daily basis (7%) also integrate digital books in their classroom instruction on a daily basis (0%). In other words, teachers who fall into this category seem to also use other digital resources apart from those provided by the Ministry of Education.

In an attempt to find out the reasons underlying this apparent lack of interest in digital books, I have analysed these results in respect to the following three questions: 6, 7 and 8.

6. How important do you think it is for children to have access to digital books?
- Very important
- Somewhat important
- Not important

Results:

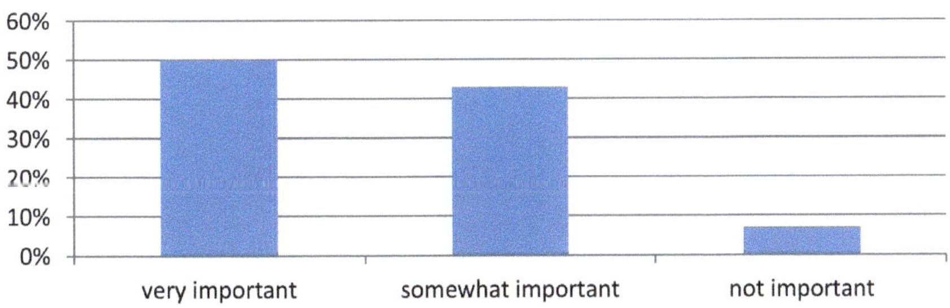

Figure 16.6: Importance of children having access to digital books.

It is noteworthy that most teachers *do* consider it important for children to have access to digital books, while only 7% of teachers seem to disregard their use in the classroom.

7. What technical drawbacks have you encountered when using digital books in the classroom? (Open-Ended Question)

Results:

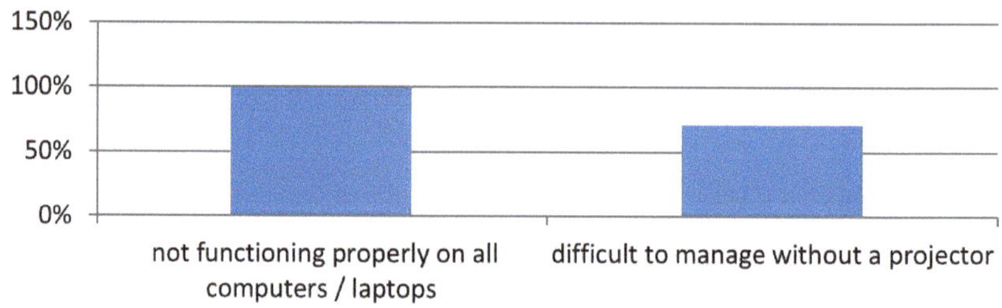

Figure 16.7: Technical drawbacks encountered.

Although teachers' answers revolved around several issues, I have chosen to analyse the two most salient ones. On one hand, all surveyed teachers pointed out the fact that digital books do not work well on all the computers. This issue may point out to the fact that there must exist a certain type of compatibility between the digital books and the computer hardware or software so that they may run properly. On the other hand, a very high percentage (71%) mentioned that digital books are difficult to manage in the absence of a projector and a screen making it difficult for children to watch the animations.

8. What are the benefits of using digital books in the classroom? (Open-Ended Question)

Results:

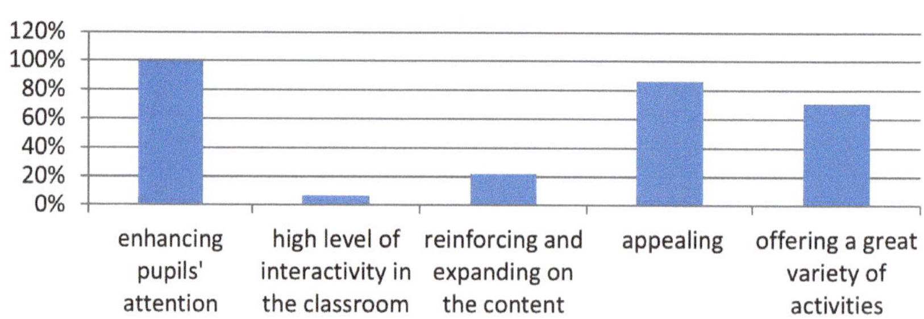

Figure 16.8: Benefits of using digital books in the classroom.

There are numerous benefits mentioned by teachers when it comes to the use of the digital books in the classroom, but the most prevalent ones centred on: enhancing students' attention (100%), being appealing (86%), and offering a great variety of activities (71%).

9. Name three features of digital books (Open Ended Question)

Results:

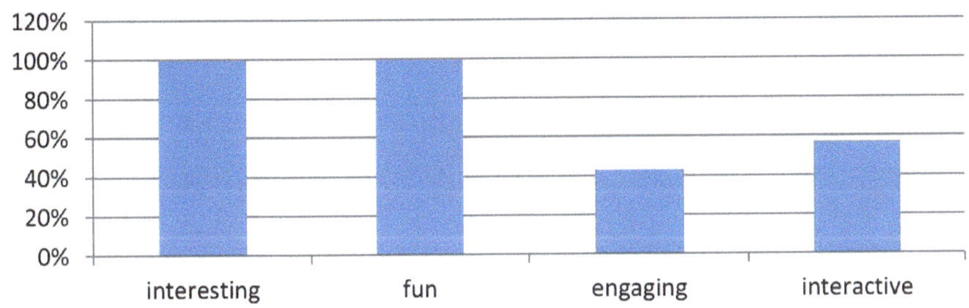

Figure 16.9: Features of digital books.

It is to be noted here that all teachers described digital books in positive terms. Digital books were valued for several aspects, but due to their relatively close percentage, I have chosen to present the first four ones: all the respondents described digital books as being interesting and fun (100%), but they also valued them for their interactivity (57%) and attractiveness (43%). All in all, they seem to be an effective way of enhancing students' motivation for the English class.

10. List the things that you do not like about the digital books (Open-ended question)

Results:

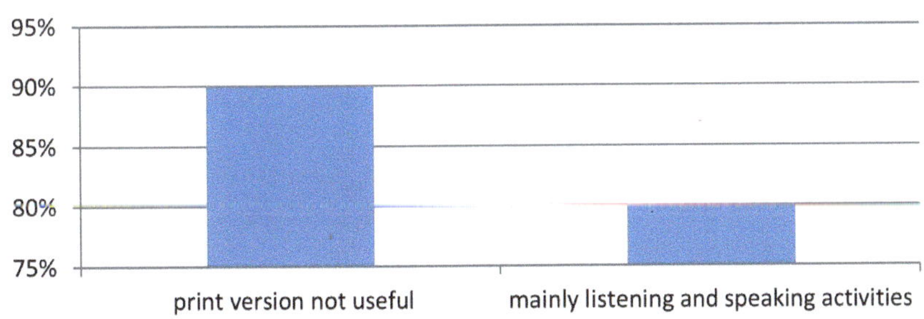

Figure 16.10: Aspects that teachers do not like about digital books.

The most salient answer to this question was the fact that the printed books are not useful at all if one cannot use the digital part as well. They are almost totally dependent on the digital version because although they have a nice format with lots of colourful images, the paper books seem to have poor content. This probably comes as a consequence of the fact that, as mentioned previously, the print version and the digital one are meant to complement each other, not to be used independently.

Another salient aspect is the fact that some teachers are dissatisfied with the fact that digital books consist mainly of listening and speaking activities with a rather limited availability of writing and reading activities. This feature may come as a consequence of the learners' early age ranging between 7-8 years old when their writing and reading skills are insufficiently developed in their own mother tongue.

16.3.3 Young learners of English and their opinions on digital books

16.3.3.1 Describing young learners

As Harmer also points out (2009:83–84), 'young learners especially those up to the age of 9 or 10, learn differently from older children, adolescents and adults'. So, what is different about teaching a foreign language to young children as opposed to teaching a language to adolescents or adult learners?

In the following lines, I will try to give an overview of the way young children learn a foreign language, what their learning needs are, and finally, I will try to answer the question whether the newly introduced digital books meet their needs or not by presenting the results of an interview carried out over 38 young learners of English.

When discussing young children's learning processes, one should take account of the specificities of their young age:

- 'They respond to meaning even if they do not understand individual words
- They often learn indirectly rather than directly – that is they take in information from all sides, learning from everything around them rather than only focusing on the precise topic they are being taught
- Their understanding comes not just from explanation, but also from what they see and hear and, crucially, have a chance to touch and interact with
- They find abstract concepts such as grammar rules difficult to grasp
- They generally display an enthusiasm for learning and a curiosity about the world around them
- They have a need for individual attention and approval from the teacher
- They are keen to talk about themselves and respond well to learning that uses themselves and their own lives as main topic in the classroom
- They have a limited attention span; unless activities are extremely engaging, they can get easily bored, losing interest after ten minutes or so' (Harmer, 2009: 82).

A good language teacher should, thus, consider all the aforementioned aspects when planning a lesson, and a good language book should enable the teacher to design age-appropriate activities and experiences by creating interest and making out of the learning experience a fun and enjoyable one because 'successful lessons and activities are those that are tuned to the learning needs of pupils, rather than to the demands of the next text-book unit, or to the interests of the teacher' (Cameron, 2002:1)

In addition, as Harmer (2009: 83) puts it, 'good teachers at this level need to provide a rich diet of learning experiences which encourage their students to get information from a variety of sources'. Among the most successful activities he mentions, one may mention puzzle-like activities, drawing, games, physical movement, songs mixing 'play and learning in an atmosphere of cheerful and supportive harmony' (Harmer, 2009:83)

Are the digital books under discussion designed to meet our young learners' needs? According to the introduction to the *Fairyland* series, it is 'a course specially designed to introduce young pupils to the English language. Young learners will be captivated by the adventures of *Woody* and *Frosty* as they enter the Magic Forest and meet *Erlina*, *Willow* and *Alvin*. Through an array of activities such as stories, songs, games and craftwork, the pupils will be introduced to the English alphabet, theme-oriented vocabulary and some simple structures. In this way, the pupils will find the learning process enjoyable as they embark on their journey to discover the English language' (Dooley, 2014: 6).

In other words, the digital books under discussion seem to offer a variety of theme-based engaging lessons focussing on age-specific activities meant to create and maintain children's interest in the English class. But, do they reach their purpose? Do children enjoy them? The following section is meant to shed some light on their effect on children.

16.3.3.2 Young learners' opinions on digital books

As mentioned previously, an informal interview was conducted with 38 first- and second-graders in schools that do have the resources and do use the newly introduced digital books. Due to the pupils' very young age, children were asked general open-ended questions, such as 'Do you like English classes? What do you like about them? What activities do you like most?' etc. The pupils' answers were, thus, analysed mainly using qualitative methods.

Students' comments on the activities in the digital books were highly positive. All learners loved classes involving activities from the digital books and valued them as being much more fun than those classes not involving them.

An overwhelming majority of pupils said that the most interesting lessons were the ones when the teacher projected short videos on a screen or when they were able to watch them on a laptop.

Concerning the stories in the Fairyland series, most children mentioned they had lots of fun watching and listening to them and seemed very motivated and engaged.

Regarding the characters in the stories, most children said they were nice and amusing and seemed to be absolutely captivated by the adventures of the magic characters: *Willow* 'a beautiful, talking tree that stands proudly in the Magic Forest', *Erlina* who 'flies around the Magic Forest, helping all the flowers and animals there', Alvin 'a friendly, funny leprechaun from Ireland', *Frosty* 'a funny little snowman, brought to life by magic' and *Woody* 'a clever little wooden puppet' who 'lives in a tree house in the Magic Forest' (http://www.expresspublishing.co.uk/elt/fairyland/meet_the_characters.html).

Children were also highly enthusiastic about the songs and chants, especially those that also involved some kind of movement accompanying the lyrics (dancing, clapping hands, waving hands in the air, etc). Some of them even unsolicitedly started singing some of the songs learnt during the English class, which only proved once again how much they enjoyed them.

In addition, most pupils mentioned they absolutely loved the games, especially those that involved miming and crafting. What was particularly interesting about children's comments was the fact that most children happily, but innocently thought that the English class was only about playing:

e.g., 1) 'We only play games during the English class'
2) 'Our teacher is the best because she allows us to play games' (*my* translations)

It, therefore, results that the digital books under discussion focus on the interactive activities that promote learning through play, making out of the English class a pleasant and memorable experience.

16.4 Conclusions

These results raise a number of significant issues concerning the use of digital books in the context of English language teaching and learning.

Children absolutely love classes in which teachers integrate the activities from the digital books. However, the digital version is used rather rarely. This does not come as a consequence of the fact that teachers do not find it appropriate or useful for the young learners, but either as a consequence of teachers' lack of resources (educational technology) or of knowledge on how to use it.

Moreover, if not used together with the digital version, the print version seems to be unusable and useless. And, since 36% of teachers say that they rarely use the digital version and 50% admit not having used it all, one may draw the conclusion that for an overwhelming number of pupils, digital books do not achieve their end, i.e., supporting learning by increasing children's interest and understanding through powerful media resources. Moreover, the teachers' impossibility to use the digital version may even make it more difficult for them to prepare, plan and deliver the lessons since they cannot truly rely on the paper book.

All in all, although digital books have a great potential to be motivating and immensely useful for the young learners, their introduction has not led to much improvement in our educational system so far. On the contrary, if measures, such as equipping *all* schools with the necessary educational technologies and organising training sessions for teachers to improve their digital skills are not taken in the near future, the whole national programme of introducing digital books in the classroom instruction will run the risk of becoming obsolete.

References

Cameron, L. (2002). Teaching Languages to Young Learners. UK: Cambridge University Press.

Dolley . & Evan V. (2014). Comunicare in limba moderna 1 - limba engleza. Fairyland 1 A. Teacher's Book, UK: Express Publishing

Pim, C. 2013. Emerging Technologies, Emerging Minds: Digital Innovations within the Primary Sector. In G. Motteram (ed) Innovations in Learning Technologies for English Language Teaching (pp15-42). London: British Council,

Fullan, M. (2007). The New Meaning of Educational Change. Fourth Edition, New York: Teachers College Press, Columbia University

Fullan, M. (2013). The New Pedagogy: Students and Teachers as Learning Partners. In Learning Landscapes. Teaching and Learning in the Digital World: Possibilities and Challenges, Spring 2013, Vol 6, no 2, Canada: Learn. Available athttp://www.learninglandscapes.ca/images/documents/ll-no12-vfinal-lr-links.pdf [accessed February 2015]

Harmer. J. (2009). The Practice of English Language Teaching (Fourth Edition). England: Pearson Longman

Motteram. G. (2013). Developing and Extending our Understanding of Language Learning and Technology. In G. Motteram (ed). Innovations in Learning Technologies for English Language Teaching (pp177-191). London: British Council.

Wong, L.L.C. (2013). Technological Innovation and Teacher Change: IT in Teacher Professional Development. In K. Hyland & L.L.C. Wong (eds). Innovation and Change in English Language Education (pp 248-262). Oxon: Routledge.

http://www.mediafax.ro/social/pricopie-vrea-tablete-manuale-digitale-si-catalog-electronic-in-toate-scolile-13440664 [accessed February 2015]

http://www.edu.ro/index.php/pressrel/22251?theme=print [accessed February 2015]

http://www.expresspublishing.co.uk/elt/fairyland/meet_the_characters.html [accessed April 2015]

Simona Olaru-Poşiar

17 Training and Development in the Digital Era

17.1 Introduction

Training is a process through which individuals are helped to learn a skill or technique. The skill may be manual, as in using a keyboard or rather intellectual, such as negotiating a contract. There is often an end-point, perhaps the achievement of a specific data-entry speed. Progress in today's digital era places emphasis on the growth of the individual, relating to acquiring a broad range of planned activities and experience that is most commonly acquired through the extensive use of a computer or other means of modern technology. The Internet has far-reaching implications for the availability of information, for education. It is changing the way we work and creating new businesses that support technology. At the same time, technology and the Internet also provide new techniques for trainers to use in the process of training itself. However, this can affect interpersonal communication.

The basis for most training remains the traditional training process system. This comprises of four main steps, such as identifying training and learning needs, devising a learning plan, delivering training, and evaluating the outcomes.

At the basis of our education, our self-progress lies on learning or better said, self-learning. Learning takes place when an individual has understood and internalised new information and/or has developed a new skill as a result of experience. Evidence that learning has taken place may be inferred from a change in an individual's behaviour. Learning is an active process, which may occur socially, systematically or experientially. All development is self-development that is people can teach, train and coach you, but nobody can learn for you. Learning is a 'do it yourself' activity. Information technology facilitates self-learning as never before, since it has broadened and deepened the available range of methods and media through which learning may take place. Harrison's (2005, 269–73) 'learning event' is based on the training process system. She defines the 'learning event' as 'any learning activity that is formally designed in order to achieve specified learning objectives'. (Marchington & Wilkinson, 2005, 242).

This typically involves the following: establishing needs, 'agreeing the overall purpose and objectives, 'identifying the profile of the intended learning population', selecting strategy and agreeing on direction and managementselecting learners and producing a detailed specification, confirming the strategy and designing of the event; delivery'(Marchington & Wilkinson 2005, 242), monitoring, and evaluation.

There are clear advantages in using structured and sequential models for analysing the training process. Whereas training is needed in the shorter term, to carry out tasks that are needed now, the term 'development' refers to the broader landscape. It relates to the future, to the longer-term development of people throughout their careers, providing them with the kind of confidence, maturity, and stability that enables them to adopt greater responsibility.

Training produces competence, while development produces continuous psychological growth/personal development. It could be said, therefore, that training is for now, while development is for the future.

17.2 The Theory of Motivation and the Protheus Effect

Motivation is the key factor in a successful training and development. While there is no universal definition of motivation, it is generally accepted to be the willingness to apply one's effort towards the achievement of a goal that satisfies an individual need. It is a natural human response to a stimulus. The response involves action designed to satisfy a need or attain a particular goal. Learning motivation can be defined as one's willingness to apply one's efforts towards the individual's long-term goals, his development, while all his needs are satisfied, especially the individual's needs of achievement, as we will explain in the following paragraphs. There is no such thing as an unmotivated person and it is motivation that triggers behaviour. One's behaviour is closely linked to one's development. Training leads to development and development leads to an individual's behaviour. Going further that thinking to a simple training that we conduct, our method as trainers, the environment, the devices that are used, the up-to-date information, all help to a better development of an individual, and moreover, to the individual's behaviour. We can relate here to the theory of behaviourism:

> "Behaviourism, also known as behavioral psychology, is a theory of learning based upon the idea that all behaviors are acquired through conditioning. Conditioning occurs through interaction with the environment. Behaviorists believe that our responses to environmental stimuli shape our behaviors ".[1]

The main interest of a trainer is to achieve his or her objective by maximising the human resource. The goal is to elicit a performance that will lead to the development of the individual and the achievement of the course's purpose. Usually, a motivated sustained workforce is a high-performing workforce. When the student does not have a proper environment, but moreover, when he lacks the proper means of information, the proper devices for a proper training to take place, the work he produces is rather poor and the outcome can be analysed in his later development and education. His behaviour is linked to his education and development.

1 http://psychology.about.com/od/behavioralpsychology/f/behaviorism.htm, accessed [27.04.2015]

Motivation is a constant factor in human behaviour and it cannot be switched off. Even if one is to drop to sleep during a course, a psychologist would rightly assume that it is what you were motivated to do. The trainer's aim should, therefore, be to change employees' motivations from what they are to higher standards in order for the student to progress, not regress, and in order to develop to his maximum potential:

> "Motivation leads to performance and the degree to which a knowledgeable and skilled individual will apply his/her best effort to a task is determined by the degree to which he/her is motivated" (Bauer & Kenton, 2005)

Additionally, it is important to bear in mind the element of individual differences. People have their own unique attitudes to work, to the university/education, to the trainer; they come from different cultures and relate differently to the trainer that is himself/herself an individual and they are not all turned on by the same motivators or possible types of outcome/development.

It is very important to underline that when we refer to trainers, we refer to teachers, professors and all the persons qualified, skilled to coach/train/teach a proper course regardless the domain, to transfer information from the receiver-trainer to a recipient of nformation-student, trainee, through up-to-date, efficient means whether we speak of the Internet or face-to-face teacher-student interaction with the goal of the trainee's development professional and implicitly personal–global development.

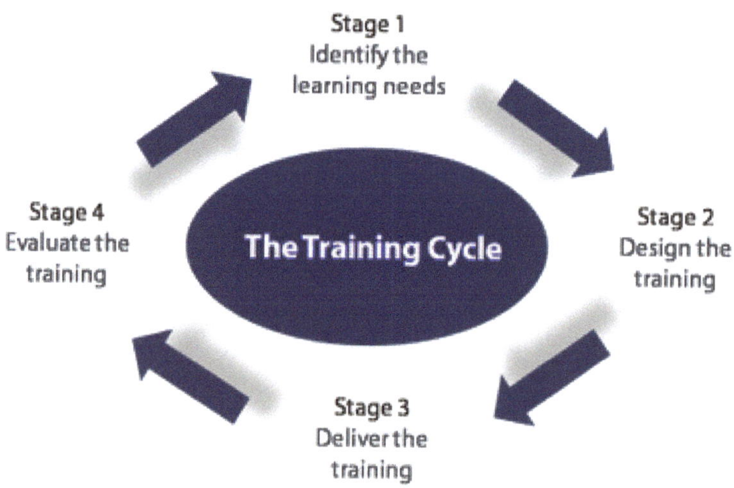

Figure 17.1: The systematic training cycle.

We should, when referring to training, take into consideration the systematic training circle. The latter is made up of four interdependent stages (Tyson & York, 1996).

The model has been successfully used in training within the public sector for many years and is regarded as a sound basis for cost-effective training. The two researchers argue that the interdependence of the components is crucial. The malfunction or neglect of any one of them inevitably affects the others or the total system. Thus, if course analysis has not defined the criteria for effective performance, training needs cannot be identified by performance appraisal. If needs have not been properly identified, it is not possible to design and provide needs-related training to asses ultimate effectiveness in terms of subsequent work performance.

Motivation lies at the basis of any type of training and development. Research has shown that in an organised learning situation, such as a training course, the amount of learning is determined by three factors: the personal characteristic of the learner, such as the motivation to learn and the intellectual ability to understand what is being said and done; the effectiveness of the trainer in terms of his or her knowledge and competence as a trainer, and the nature of the methods and media through which he or she delivers the training; the physical situation in which the training takes place. For education, training and development to be successful, learning must take place. The circumstances in which learning is acquired have been the focus of academic interest for at least one hundred years. The ultimate purpose of training must, nevertheless, be the improvement of academic performance. Individuals, however, may regard training as one of the means by which they can improve themselves in order to enhance their future careers. Viewed in this way, training may be seen as a developmental driver. Developing a training course involves professional expertise of the trainer both pedagogical and psychological, without taking into account the domain taught. Today's digital means such as the Internet, the devices that must be always up-to-date with the needs of the students' computers, video devices, help training and develop the individuals' maximum potential. In order, however, to speak about competences, training and development, we must identify the five fundamental human needs, as described in human psychology:

> "Created by psychologist Abraham Maslow, the hierarchy is often displayed as a pyramid, with the most basic needs at the bottom and more complex needs at the peak. The four lowest-level needs are what Maslow referred to as D-needs (or deficiency needs). These needs are due to a lack of something and need to be satisfied in order to avoid unpleasant feelings and to move on to higher level needs. The uppermost needs in the hierarchy are referred to B-needs (being needs or growth needs) and involve the desire to grow as an individual and fulfil one's own potential"[2].

The scientist believed that all of us are motivated by needs. Maslow Hierarchy of Needs describes how rudimentary needs must be fulfilled before more complex needs or desires are contemplated. There are as mentioned above, five stages: biological and physical needs, safety needs, belonging and love needs, esteem needs, and at the top,

2 www.psychology.about.com/theoriesofpersonalities/malows-needs-hierarchy, accessed [27.09.15]

self-actualization. What is new in today's digital era is that at the base of the pyramid lie the so-called Wi-Fi needs, adding another level at the ground of our fundamental needs in life. We cannot work properly without our avatar personality, our Facebook or Twitter image. We need to be liked, we need appreciation via the Internet, we need to be shared and posted. We gather friends from people we haven't personally met and give a clear depiction of our location. Everybody knows the portrait of our perfect self. The *Dorian Gray* we draw is flawless. This is also known as the *Protheus Effect.*

> "Whether it's an animated alter ego in a game or online community, or a two-dimensional Facebook profile picture or Twitter, an avatar, for lack of a better explanation, is our incarnation on the Internet – the virtual image we expose every day. Our avatar is how the online world sees us. It's also how we see ourselves"[3].

The only exception to Oscar Wilde's portrait is that we share it, we don't keep it locked. We portray a reality that is too much digital and less human, leaving the interpersonal down-to-earth reality an obsolete detail. University interaction, although professional tend to follow the trend. The World Wide Web doesn't distinguish between racial or gender identities and offers a permanent curiosity on exploring and discovering new worlds.

Distance education, e-learning seems to be the quickest method in the digital era. The bottom traditional four levels of needs are described as deficiency motivators by Maslow, because when we fulfil these needs, we satisfy a need for something we do not have. The top level need is characterised as a growth motivator because self-actualization means that a person is realizing their personal potential, seeking peak experiences for personal growth. As facilitators and instructors, it is important to understand why the students are in the class. Students are individuals at different levels within Maslow's hierarchy of needs, because the motivations for being there are unique to each situation. By addressing the needs, a professor should make every effort to see that the basic needs are fulfilled. The class environment should be physiologically and mentally healthy. To properly address safety needs, the training environment must be perceived as free of threats and dangers. To fulfil the social needs of the next level, the teacher or trainer must create a feeling of acceptance and reinforce positive class dynamics. The esteem needs of students are fulfilled by recognition of achievements and regular progress updates. The final stage of self-actualization can be addressed by having the students create meaningful projects that enable them to use their innovation and creativity. There is no substitute for experience.

3 http://mediashift.org/2011/09/our-avatars-ourselves249, accessed [27.04.15]

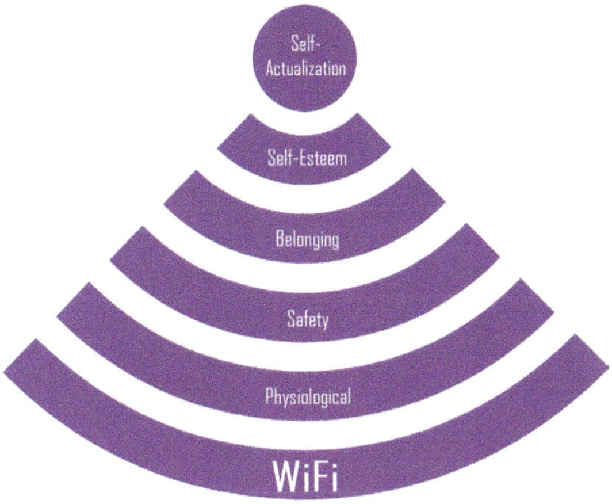

Figure 17.2: Maslow's pyramid of needs reinterpreted in the Digital Era.

In contradiction to Maslow's classification, there were other three categories of need added by Paul Alderfer in 1972, a reinterpretation of Maslow's pyramid of needs. Essentially, what the latter did was to draw a parallel between Maslow's five categories and three categories of his own, which he referred to as existence, relatedness, and growth.

Existence needs are concerned with physiological, safety and security needs and cover all needs of a material nature, which are necessary for human survival. Relatedness needs mean those for love, esteem and belongingness, while growth needs are represented by achievement, recognition and the realisation of potential: what Maslow called self-actualisation. This is also known as the ERG theory. In the 20th century, two main approaches to motivation were proposed, theories that are followed and are underlined in training today as well. The first is based on content theories and state that people are said to have needs which may (to some degree) be satisfied by the factors that make up the task. Second are process theories, in which our experience has taught us that a certain behaviour produces particular outcomes and we do in the expectation of achieving desired outcomes. In this way, motivation is generated through a mental process, rather than as a response to particular environmental factors.

17.3 Digital Learning in Traditional universities versus Non-traditional universities (Technology-based Learning)

Online web-based universities emerged with the development of computer conferencing systems and the World Wide Web; many new online universities have been established in the past years. These universities are coming into existence specifically to use new web technologies that support learning independent of time, location, but allow for students to study together. They offer opportunities for students to learn through asynchronous interaction with each other and a faculty member. A classroom environment with student and faculty interaction is created, but students are not all in the classroom at the same time. Online universities define their competitive market advantage based on the convenience of electronic computer-based access they provide to specific programs. Unlike the national distance learning universities, which have a historical tradition tied to correspondence study and the post office, these new universities focus on the use of new technologies to provide not only improved access, but also improved interaction between and among students. While their numbers are relatively few and their structure is evolving rapidly, following is a list of exclusively online universities and organizations with the name of university currently available on the Web, for example, Athena University, California Coast, University American, Coastline University, Commonwealth Open University, Cyber State University, Greenleaf University, Kennedy Western University, International University, Open University, Southern California, University of Professional Studies, Virtual Online University. (Hanna, 1998, 81)

> "Technology Based Universities Keegan categorizes *distance education* universities as originating from two distinct traditions. The first of these traditions is correspondence study, and the second is the extension of traditional classrooms to new locations through the use of new technologies such as satellite, broadcast television, cable television, and more recently, compressed video and desktop video. More recently, a third category of institution has emerged that does not neatly fall into either of these traditions. Using asynchronous learning and taking advantage of new computer mediated conferencing systems and the emergence of the World Wide Web, *online universities* offer a third model organized around a technology approach. The *distance education/technology-based universities* are all organized around a technology-based approach to learning that seeks to minimize the physical separation of the learner from the instructor or from other learners. They also tend to be more adult and workforce oriented, although the large national universities enrol substantial numbers of traditional college-age students largely due to the incapability of traditional universities, especially in countries with rapidly growing populations" (Hanna, 1998, 77)

Learning and instruction will become increasingly interdisciplinary, due to the intermingling of the professional and personal sides of the learning process. Academic departments will be encouraged administratively and driven economically to reformat and reorganize trainings, programs, and structures to respond to increasingly market-orientated students. Technology support units that in traditional universities have been concerned only with improvements in on-campus. However, instruction will find that their work intersects with continuing education units whose role has been to extend access to the programs through the use of technology.

> "The recent developments of the World Wide Web, digital satellite technology, and new applications of virtual reality to build simulated learning environments, are predicted to have particu-

> larly dramatic effects upon learning environments at all levels. Universities are experimenting with improving accessibility to existing programs, designing new programs to take advantage of these emerging technologies, and are marketing their programs to new audiences and in new ways. Corporations are also engaged in experimentation and have formed both new organizations internal to the corporation and brand new alliances with universities to promote learning using technology. Completely new models for universities are also being developed to respond to the opportunities created by a growing worldwide market for learning and new technologies. The result is a dynamic competitive environment among traditional universities that are adapting learning processes and administrative procedures, alternative non-traditional universities that are adapting technologies to better serve their existing primarily adult constituencies, and new universities that are being formed around the promise of virtual environments" (Hanna, 1998, 67)

Technologies are and will be used in creative ways to further erode the separation of students from each other, from their teachers, and from content relevant to the needs and interests of the student. As all of this occurs, the truly global nature of the educational marketplace will become increasingly clear, just as it has become apparent in this decade that the market for higher education is no longer singularly local. It will also become clearer that the impact of technology is not to create mass markets for learning, but to create options that are more and more customized for individual learners in organized patterns of inquiry.

The debate is between formal and informal learning experience, identity and avatar psychology, enhanced immediate synchronous e-learning interactions, experiential and social learning, virtual teamwork, formation of virtual communities, effective behaviour-changing learning experiences.

> "Digital literacy involves more than the mere ability to use software or operate a digital device; it includes a large variety of complex cognitive, motor, sociological, and emotional skills, which users need in order to function effectively in digital environments. The tasks required in this context include, for example, "reading" instructions from graphical displays in user interfaces; using digital reproduction to create new, meaningful materials from existing ones; constructing knowledge from a nonlinear, hyper textual navigation; evaluating the quality and validity of information; and have a mature and realistic understanding of quoting that prevail in the cyberspace. This newly emerging concept of *digital literacy* may be used as a measure of the quality of learners' work in digital environments, and provide scholars and developers with a more effective means of communication in designing better user-oriented environments. It includes photo-visual literacy; reproduction literacy; branching literacy; information literacy; and socio-emotional literacy". (Yoram, 2004)

17.4 E-learning

Computer simulation can be viewed as a form of e-learning. This is a wide term covering opportunities that are proposed by the electronic age and deserves separate attention and mentioning. It can, of course, take place within the university or off, when learning at home or at home via the Internet when taking an online course or within an online university study. The term includes both PC-based learning and web-based learning: 'E-learning is the use of electronic educational technology in teaching and learning'.

> "Bloom's taxonomy is a classification system used to define and distinguish different levels of human cognition—i.e., thinking, learning, and understanding. Educators have typically used Bloom's taxonomy to inform or guide the development of assessments (tests and other evaluations of student learning), curriculum (units, lessons, projects, and other learning activities), and instructional methods such as questioning strategies. Critics of the original taxonomy have questioned whether human cognition can be divided into distinct categories, particularly sequential or hierarchical categories. Others embrace the utility of the classification system, while still recognizing that it does not—and cannot—represent human thought or learning in all their complexity and sophistication. Most criticism is focused less on the system itself and more on the ways in which educators interpret and use the taxonomy. For example, trainers may view the system as linear prescription, believing that students must first begin with remembering, move on to understanding, and proceed through the levels to creating. Other educators may place too much emphasis on the importance higher-order thinking—at the expense of lower-order skills—despite the fact that acquiring a strong foundation of knowledge, information, and facts is essential in the application of higher-level thinking skills. Some educators have even proposed an alternative formulation, suggesting that the taxonomy should be reversed because higher-level thinking skills require that students both remember and understand underlying concepts first. Others suggest that the taxonomy should be interpreted as a non-hierarchical continuum in which no one form of cognition is more or less important".[4]

The figure above pictures the taxonomy linked to modern day means of communication, such as blogging, Skype communication, video conferencing, digital means used by online universities, and the online environment in general.

There are occasions when the media the trainer uses is determined simply by what is available in the training facility, and sometimes, therefore, this has to make do. Having said that, each of the media that are mentioned above, lend themselves to particular subject matter.

> "Information and communication technology (ICT) in education, Ed Tech, learning technology, multimedia learning, technology-enhanced learning (TEL), computer-based instruction (CBI), computer managed instruction, computer-based training (CBT), computer-assisted instruction or computer-aided instruction (CAI), internet-based training (IBT), flexible learning, web-based training (WBT), online education, online learning, virtual education, virtual learning environments (VLE; which are also called learning platforms), m-learning, and digital education".[5]

4 http://edglossary.org/blooms-taxonomy/, accesed [20.05.2015]
5 https://en.wikipedia.org/wiki/Educational_technology, accessed [27.04.2015]

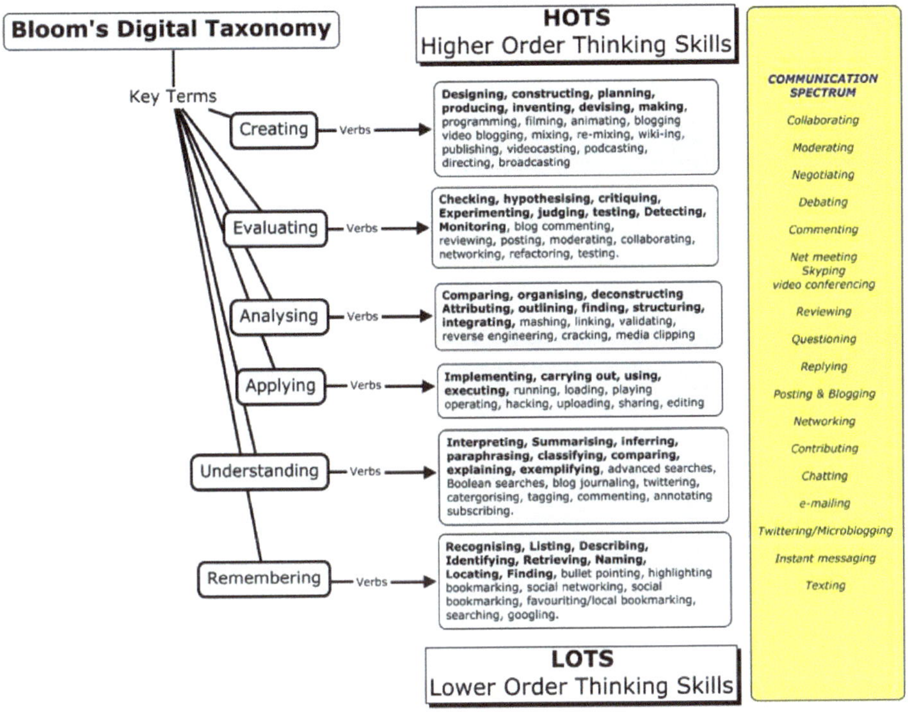

Figure 17.3: Bloom's Digital Taxonomy.

Information technology has extended the range of training media, especially in the area of visual aids. These include PowerPoint and DVDs. Other visual aids include overhead projector (OHP), whiteboard, flip chart, videos and films. Written material, such as handouts, is also a form of medium. Finally, one's voice as a medium must not be overlooked, since it is a means by which one is conveying information:

> "While still widely used, Bloom's taxonomy is gradually being supplemented—and may perhaps even supplanted one day—by new insights into the workings of human thought and learning made possible by advances in brain imaging and cognitive science. Still, it is likely, given its logical simplicity and utility, that Bloom's taxonomy will continue to be widely used by trainers".[6]

Searching the web can also unearth other online training opportunities. This is a rapidly developing area when reading the current literature and networking will help to keep one up-to-date. Most commentators in this area agree that e-learning that also has a tutorial support, perhaps via e-mail, is more successful than other approaches.

6 http://edglossary.org/blooms-taxonomy/, accesed [10.05.2015]

Web information that might be useful for a student, include access to World Wide Webuniversities' resources, mailing lists, forums, conferences and of course, online.

Meanwhile, Table 17.1. lists the most commonly used media and the purposes to which they are suitable.

Table 17.1: Commonly used visual training aids

Medium	Usage
PowerPoint	Everything is prepared on the computer. The package usually includes instructions for use. It is usually to run off copies of the PowerPoint frames to hand out to the trainees as *aides memoire*. Within the medium, the trainer can use colour and animation.
OHP	Suitable for presenting material in a clear fashion, explaining and discussing as the course develops.
Whiteboard and flip chart	Handy for summarizing findings, trainee's answers to the trainer's questions, and for laying out the main points of a talk.
DVD, video, and film	Films and video materials created specifically for training purposes in which actors adopt roles in case studies and problem solving situations.

17.5 Conclusions

We have looked at the steps of the training cycle and examined the management of training and development in the digital era. It is the changes in the environment in which a student develops that create the need for training.

> "In today's digital, technological and social environment, important transformations are underway in terms of how we live and work. We refer to contemporary times as the digital era or knowledge based society, characterised by the diffusion of information and communications technologies and the increasing demand for new educational approaches and pedagogies that foster lifelong learning. In the higher education arena, there are shifts in the views of what education is for, with a growing emphasis on the need to enable and support not only the acquisition of knowledge and information, but also to develop the skills and resources necessary to engage with social and technological change, and to continue learning throughout life. We are witnessing the rapid expansion and proliferation of technologies that are less about narrowcasting, and more focussed on creating communities in which people come together to collaborate, learn and build knowledge" (McLoughlin & Lee, 2007, 664).

Training needs can be identified within an academic institution, at the individual level. We have to consider not only what may require training, but whether that is both important for the future job holder and likely to be recognised or rewarded within the future organisation where the individual will be integrated. Training needs are established by examining the gap between the performance that is sought and the performance that is currently being achieved. A wide variety of sources is available to help determine both the desired performance and the current performance. Competency frameworks can be particularly useful.

> "To illustrate the power of social software to support learner-centred pedagogies and to support courses in General Psychology, teachers at on-line universities, that use social software tools, should host weekly informal discussions with students following each week's lectures. During these discussions, students should be able to seek clarification on the course material and talk about it in greater depth, as well as to discuss issues not covered during the lecture. The discussions must be recorded and made available to other members of the class as a series of podcasts, or through Skype for example. In this way, the podcasts are about course content (meta-cognitive) rather than simply being recordings of the course content itself (transmission of content). All students in the cohort are welcome to submit questions in advance of the discussion via email; these answers, as well as those asked by students who attend in person, should be answered during the discussion".[7]

In formulating plans for training and development, it is important to examine the internal resources available, the external resources and the relevance of qualifications. We can select from a wide variety of techniques and opportunities, and should never restrict our concept of training and development to training courses alone. Individuals have preferences for the ways in which they learn, learning styles. The choice of training activities should take this into account.

Online universities provide the future in what digital learning is concerned. They offer a fast and cost-effective way of learning. To complete the learning cycle, we emphasised the value of evaluating training, considered some of the practical obstacles, suggesting examples that could help structure of evaluation. Without new and broader horizons and without taking advantage of new information and ways to improve training, development cannot take place. Therefore, training can be enhanced by technological means and is vital for progress.

> "Research in the past decade has shown that computer technology is an effective means for widening educational opportunities, but most teachers neither use technology as an instructional delivery system nor integrate technology into their curriculum. This qualitative study examined the classroom practice of 30 "tech-savvy" teachers who used computer technology in their instruction, how much they used it, the obstacles they had to overcome to succeed in its use, and their general issues and concerns regarding technology. Participants were volunteers from two elementary schools, one middle school, and one high school. All identified by their schools as being proficient with technology. The study found that the teachers were highly educated and skilled with technology, were innovative and adept at overcoming obstacles, but that they did not integrate technology on a consistent basis as both a teaching and learning tool. Two key issues were that their students did not have enough time at computers, and that teachers needed

7 http://www.thefreelibrary.com/Towardtechnologyintegrationintheschoolswhyitisn'thappening, accessed [27.04.2015]

extra planning time for technology lessons. Other concerns were out-dated hardware, lack of appropriate software, technical difficulties, and student skill levels. Results suggest that schools have not yet achieved true technology integration. There are implications for teachers, administrators, and teacher educators".[8]

Technological change is relentless. Training needs created in computer skills, in advanced technical skills are widespread and substantial. The ability to develop and exploit software opportunities is critical for self-progress and such new software means the need for training. The Internet has far-reaching implications for the availability of information, for education. It is changing the way we work, communicate and learn. At the same time, technology and the Internet also provide new techniques for trainers to use in the process of training itself.

References

Alderfer, C. P. (1972). Existence, relatedness and growth: human needs in organisational settings. New York: Free Press

Bauer, J. and J. Kenton (2005). Toward Technology Integration in the Schools: Why It Isn't Happening. Norfolk, VA: Society for Information Technology & Teacher Education -13(4), pp 519-546,

Bramley, P.. 2003. Evaluating Training.UK: CIPD Publishing.

Daniel, J. (1996). Mega-Universities and Knowledge Media: Technology Strategies for Higher Education. London: Biddles Ltd

Gazzaniga, M.(2010). Psychological Science. New York: W.W. Norton & Company

Hanna, D. E. (1998). Higher Education in an Era of Digital Competition: Emerging Organizational Models JALN Volume 2, Issue 1 – March 98 p.77

Harrison, R. (2005). Learning and Development. London: Chartered Institute of Personnel and Development.

Hoffman, E. (1988). The Right to be Human: A Biography of Abraham Maslow. New York: St. Martin's Press.

Horton, W.. (2000). Designing Web-Based Training. New York: Wiley & Sons.

Marchington, M. and A. Wilkinson (2005).Human Resouce Management at Work London: Chartered Institute of Personnel Practice

McLoughlin, C. & M. J. W. Lee, (2007). Social software and participatory learning: Pedagogical choices with technology affordances in the Web 2.0 era. In ICT: Providing choices for learners and learning. Proceedings ascilite Singapore .Price, K. M. and K. L. Nelson (2011). Planning effective instruction. Diversity responsive methods and management. Belmont: Wadsworth

Tapscott, D. (1996). The Digital Economy: Promise and Peril in the Age of Networked Intelligence. New York: McGraw-Hill

Tapscott, D.(1998). Growing up digital. New York: McGraw-Hill

Tyson, S. and A. York.(1996). Human resource management. Oxford: Butterworth-Heinemann.

Yoram, E.-A. (2004), Digital Literacy: a conceptual framework for survival skills in the digital era, Journal of Educational Multimedia and Hypermedia, March 22nd, Malaysia:The University of Adelaide, digital library

8 https://en.wikipedia.org/wiki/Educational_technology accesed [27.04.2015]

Internet links

www.cipd.co.uk accessed [27.04.2015]
http://edglossary.org/blooms-taxonomy/ accesed [20.05.2015]
http://www.thefreelibrary.com/Toward+technology+integration+in+the+schools%3A+why+it+isn't +happening.-a0138483291 accessed [27.04.2015]
http://mediashift.org/2011/09/our-avatars-ourselves249/ accessed [27.04.15]
http://psychology.about.com/od/behavioralpsychology/f/behaviorism.htm accessed [27.04.2015]
www.psychology.about.com/theoriesofpersonalities/malows-needs-hierarchy accessed [27.04.15]
https://en.wikipedia.org/wiki/Educational_technology accessed [27.04.2

Valentina Mureșan

18 Developing Communication Skills in Romania in the Digital Era

18.1 Introduction

The starting point for my study is Don Tapscott's remark (2008:126) about the increasing number of school leavers in the USA, which he explains as the refusal of this generation of learners to accept 'the old broadcast product'. The group of learners who are prone to give up being part of the educational system belong to the newer generation of students, born after 1977, that is, members of the 'Net Generation' or generation Y (born between 1977 and 1997) and of the 'Generation Next' or generation Z (after 1988) (Tapscott, 2008:16). To support his claim for rethinking modern teaching, Tapscott uses the findings of a 2006 report, where students justified their disinterest by saying that 'classes were either not interesting or just plain boring' (2008:126). Sadly, the above-mentioned situation is very similar to what we experience in Romania, where the passing rate for the final secondary school exam (the Baccalaureate) has dropped in the last few years.

Starting from the postmodern learner needs and the demands of the digital age society, I propose to take a closer look at foreign language teaching in the secondary education system, with a focus on the importance of developing a communicative competence.

For a more accurate description of the context of secondary school teaching in the local community, I rely on the remarks and comments of participants in the *Blended Learning* programme (a European funded training project for secondary education teachers). During 2008–2012, I was a co-trainer in this project, responsible for the *Web 2.0 in Education* training sessions, together with colleagues Phd. Mihaela Tilincă, Tatiana Cărăbaș, and Melinda Moldoveanu. Furthermore, for the discussion on how foreign language teaching fosters the development of communicative skills, I rely on my doctoral research, which I conducted in a local secondary school in 2009.

18.2 Are Teachers Prepared for Digital Age Teaching/Learning?

The *Blended Learning* teacher-training programme covered a twelve-month period for two series of seventy teachers from the Banat region, further divided into three smaller groups per series. Within this project, the topic of *Web 2.0 in Education* was one of six face-to-face training sessions, the remaining topics being: *Research in Education, Career Management, Project Management, Evaluation and Course Design.* The training sessions were followed by four hours of mentoring sessions and two unconventional (non-classroom) events each month.

It is in this context that the participating teachers were asked about the possible meaning of web 2.0 and the majority was not familiar either with the concept or with most 2.0 tools. At this point, I was taken by surprise, as I had anticipated a large number of participants to be using technology on a regular basis, in all the stages of the teaching/learning process, from lesson preparation to classroom activities and later for designing homework tasks to encourage students to develop new skills. The exceptions were teachers of informatics technology and some isolated cases, who used the computer regularly to prepare for their lessons or for professional development through online courses. The project aimed at selecting a very diverse group of participants, with teachers involved in teaching from elementary to secondary school, ranging in age between 20 and 60 years, teaching different subjects and functioning in schools from both the urban (the majority) and rural environment. The surprise was that young teachers, in their twenties or thirties, were not among the frequent computer users for school activity.

Together with the other trainers, we opened the discussion about the relevance of teaching in the Digital Age and introduced the participating teachers to the expectations of a generation that is familiar with technology and has to develop new skills in order to cope with the demands of a very dynamic job market. We invited teachers to reflect upon the expectations of the new learner and the challenge of being relevant for this new generation, by exposing them to a series of online materials, starting from the popular YouTube video *A Vision of Students Today,* created by anthropologist Michael Wesch of Kansas State University in 2007 to inspire teachers to reconsider learners' needs (Tapscott, 2008:121–122). Later, we added Sir Ken Robinson's invitation to change paradigms in education (Robinson, 2008), or Sugata Mitra's *School in the cloud* concept (a new approach to schooling, where learners learn best from one another, by discovering together the solution to learning tasks), on TED Talk (Mitra, 2013).

An unexpected outcome was that although some teachers were resistant to change, more than a third accepted our invitation to explore different uses of technology and reconsider their teaching strategy in order to better 'engage' learners. Their resolutions 'to link the virtual world and its many facets with the charming world of books and language', to 'use more interactive methods', to 'keep an open mind' or 'to be a click away' [my translation] were written on self-addressed postcards, which we sent a few months later to remind them of their promises.

However, for nearly every reformed teacher, ready to adjust to the demands of a Digital Age, there was a teacher who resisted change and who believed that Net Geners 'waste' their time in front of the computer, they do not read and they do not socialize. From group discussions during the training sessions, I concluded that the problem was not so much the lack of facilities or old technology, which is slowly catching up with what happens in other countries, but the teacher's mindset (a term used in this study with the meaning of attachment to the values of/identification with a particular approach to teaching).

To conclude, the challenge of teacher trainers was to battle the old mindset and open the teachers towards instructing and equipping their students for the future and not for the past. For a clearer picture of the transition that needs to take place in postmodern teaching, I include below Tapscott's overview of the two teaching paradigms.

Table 18.1: Broadcast Learning Versus Interactive Learning (Tapscott, 2008:133)

Broadcast Learning	Interactive Learning
Teacher-centred	Learner-centred
One-size-fits-all	One-size-fits-one
Instruction: learning about	Discovery: learning to be
Individualistic learning	Collaborative learning

18.3 Why Focus on Communication Skills?

Tapscott (2008) stresses that an important part of the solution for teaching the Net generation is addressing the issue of boredom in class, and brings as evidence the first recommendation of the researchers investigating school leaving in the USA, which is to 'improve teaching and the curriculum to make it more relevant and engaging for young people' (2008:126). Tapscott continues this idea of 'engaging learners' and pleads for teachers to 'step off the stage and start listening and conversing instead of just lecturing' (2008:130), because 'kids who have grown up digital expect to talk back, to have a conversation' (2008:126).

Thus, communication and oral interaction are part of classroom expectations, but these are not just factors that favour student participation in class, but a skill that employers expect students to develop in school. Moreover, I claim that what happens in class at the level of oral communication constitutes the basis for constructing a personal model of communication as adults, so, exposure to teacher lecturing will not only affect the student's interest in that particular class, but it contributes to a faulty communication pattern embedded in the mind of the learner.

This view is supported by two recent news articles, which discuss the question of equipping students with a communicative competence. Dumbrăveanu (2006) and Dogioiu (2008) draw attention to the Romanians' low communicative skills and call it 'a problem of Romanian society with roots in the school' and 'the most serious disease of Romanian society' [my translation]. Dogioiu (2008) attributes the problem to the issue that 'Romanians talk a lot and communicate very little' [my translation] and mentions that communication practices do not rest upon dialogues, but communication rather takes the shape of a set of monologues. She advocates introducing rhetoric in the curricula as an elective subject, as she claims that the blame does not 'rest only with those who do not have this ability, but the major culprit is the educational system' [my translation].

Similarly, Dumbrăveanu (2006) discusses the results of a recent longitudinal research project, conducted by the National Institute of Education on the topic of developing communicative competences in the compulsory formal educational system, comprising an analysis of both regular classes and extracurricular activities. The findings of the above-mentioned research reveal that in the curricula, 'communicative competence often resumes to knowledge of specialised terminology, even when there is reference to expressing one's opinion; communicative competence as interaction is rarely mentioned, if mentioned at all, in the curricula'. Furthermore, if there seems to be some focus on the production and reception of oral and written messages in the context of the language classes, in teaching of Romanian language the 'routine of an academic "approach" ignores real life communication and promotes teaching an artificial language' [my translation]. Dumbrăveanu also signals that in the context of the science classes, communication only implies formulating personal interpretations of diagrams, phenomena or laws. Finally, she concludes that, in addition to problematic curricula, their interpretation and application by the teacher is also disputable, as she often observed that teachers only indicate the number of the exercise and the page, without stressing the real-life utility and applicability of the activity.

Perhaps of all school subjects foreign language classes are the context where teachers deliberately set out to build a communicative competence and students can experience the immediate real-life impact of being able to communicate effectively in a new language.

Moreover, within language exams, the communicative competence is one of the five components under scrutiny. English language exams are fashionable in Romania nowadays and overall Romanian learners achieve fairly good results; however, my experience as an oral examiner for the speaking component of Cambridge ESOL Examinations (CEFR levels A1-C1) and that of my fellow examiners is that Romanian candidates generally underperform in the speaking part, as many do not master a higher level of communicative proficiency in the English language, characterised by features such as flexibility, naturalness and precision. The popularity of English language exams can be verified by the large number of candidates: 26,000 for general and specialised English language exams in 2013, according to the representatives of British Council Romania (Gavrilă, 2013). The reasons behind the choice to sit these exams are mainly grouped around studying and career opportunities in an increasingly mobile community, but these may also extend to comprise cultural and social aspects.

Furthermore, there seems to be a demand for foreign language proficiency, as EU reports reveal that Romanians are falling a little behind the rest of Europe when it comes to learning foreign languages. Only 48% of the Romanians say they can converse in another language, as opposed to 54% of the Europeans; the foreign language that most Romanians speak is English (31%), according to Education Commissioner, Androulla Vassiliou (Solomon, 2012).

18.4 The Communicative Competence and Language Teaching

To continue the discussion regarding the role of the school in building a communicative competence, it is worth mentioning the view of a group of professionals in education, including a former minister of education (Miroiu *et al.*, 1998), who have diagnosed the Romanian educational system as one which does not revolve around building competence in all its aspects, but rather focuses on one dimension – knowledge input. Moreover, of all key competences, building the communicative competence (not in the strict paradigm of the Communicative Approach, but in a more general understanding of an educational goal, irrespective of the subjects taught) has been 'the Cinderella' of the system, with very few classroom activities to encourage communication, although, in theory, the Romanian school syllabus aims at building and developing *all* key competences (Comisia Europeană/ EACEA/ Eurydice, 2012).

In order to explain why foreign language teaching does not really address this problem, it is important here to clarify the concept of *communicative competence,* both as a term belonging to language teaching terminology and pedagogy. Here, I rely on Celce–Murcia's (2007) model of communicative competence, which includes a subset of six interrelated competences – discourse competence, interactional competence, linguistic competence, socio-cultural competence, interactional competence, formulaic competence and strategic competence. Celce–Murcia's model continues earlier efforts to understand and define the concept of communication competence – Chomsky (1965), Hymes (1971), Canale and Swain (1980), Canle (1983), Bachman (1990), and Bachman and Palmer (1996).

According to Celce–Murcia (2007), *linguistic competence* and *formulaic competence* are different entities, the former involves knowledge about the open-ended linguistic systems (phonological, lexical, morphological and syntactic) and the latter refers to pre-fabricated phrases or 'chunks', which are heavily used in everyday interaction (2007:47). The driving force, the *strategic competence* is defined as 'an available inventory of communicative, cognitive, and metacognitive strategies that allow a skilled interlocutor to negotiate meanings, resolve ambiguities, and to compensate for deficiencies in any of the other competences' (2007:44). *Discourse competence* refers to 'the selection, sequencing and arrangement of words, structures and utterances to achieve a unified spoken message' and is manifested in aspects such as cohesion, deixis, coherence, and generic structure (2007:46). *Socio-cultural competence* refers to 'the speaker's pragmatic knowledge, i.e. how to express messages appropriately within the overall social and cultural context of communication' (ibid.), including language variation in accordance to the socio-cultural norms of L2 (a second language). Finally, *interactional competence* is defined as a hands-on competence with three components – actional competence (knowledge of how to perform common speech acts), conversational competence (knowledge of the dialogic strategies in communication such as turn-taking, interrupting, opening and closing conversations, changing topics, collaborating, backchannel, etc.), and the non-verbal or paralinguistic competence (implying knowledge of kinesics, proxemics, non-verbal signals and non-linguistic utterances with interactional import) (Celce–Murcia, 2007:48–49).

Celce–Murcia (2007:51) concludes the discussion on communicative competence with the remark that the new pedagogy should be governed by a number of principles to ensure that learners are linguistically and culturally competent in the L2. These principles include: a focus on the importance of integrating language instruction with cultural and cross-cultural instruction, with the need to use materials that are well contextualised and meaningful to the learners, with objectives that include 'some type of real-world discourse', on the importance to address functional elements and not just structural elements (as Celce–Murcia views it, the challenge is to 'maintain a balance' between a focus on vocabulary and stock phrases and one on grammar and pronunciation (2007:52)), the need to address the context dynamic aspects of interaction (rhythm, intonation, body and eye movements in face-to-face communication), and also an active involvement in the development of learner strategies in order to speed up the learning process.

Here, we can look at a particular example of two local secondary school English language teachers, whose oral classroom discourse I have analysed during my doctoral research. In my study, I followed the construction of oral communication in the classroom, focussing both on its effectiveness in general and on the interactional aspects in particular, in the context of a form of discourse that is generally considered to display distinctive, characteristic features of institutional discourse (Thornbury & Slade, 2006:242).

The data reveal an imbalance between the linguistic and discourse competence of the two teachers, one teacher's oral production corresponding to what is expected of a proficient speaker, which is also reflected in the quality of her student's responses (who tend to give longer, more complex answers both at the level of discourse and linguistic form). This teacher frequently provides extended stretches of language and makes use of numerous multiword patterns considered formulaic language, that is a typical feature of naturally occurring spoken language (Celce–Murcia *et al.*, 1995). In addition to being exposed to such semi-preconstructed phrases (idioms, proverbs), her students are also accustomed to decoding finer nuances of meaning, forms of irony or sarcasm. The other teacher's classroom discourse can be characterised as less complex, using a narrower range of grammatical forms and lexical variation, which is reflected in the language his students use. He produces shorter stretches of language, mostly consisting of basic or fragmentary sentences, with a limited use of cohesive devices. Although the classroom language of the students in the second teacher's class displays the same features as that of their teacher, at the level of classroom interaction, there is evidence of more self-selected turns (students volunteer to speak without being nominated by the teacher, yet students do not own most of the turns), which places this teacher–student interaction closer to the naturally occurring conversation.

However, from the point of view of building interactive competence, neither teacher represents a model to be followed, as the interaction between each of the teachers and their students displays an asymmetrical distribution of turns, with teachers initiating most turns and with a topicalization by teacher (it is teachers who decide on the topic of the oral exchange).

The paradoxical expectation is for students to develop a conversational competence in educational contexts, where actual conversation very rarely occurs. However, Thornbury and Slade (2006:245) claim that it is possible to turn classrooms into contexts where both at the level of task design and curriculum, classroom talk is structured 'along more conversational lines', which is why, in the next section, we turn to the educational context and the adopted pedagogical approach to foreign language teaching.

18.5 Foreign Language Teaching Pedagogy in Romania

My claim is that current teacher practices have to be viewed from a historical perspective as the product of a pedagogy, which has evolved in leaps, moving through three major approaches in comparison with the numerous philosophical views and efforts to reform pedagogical practices that characterise other English speaking countries. As I discuss in the following section, Romanian foreign language pedagogy has developed at the level of generally adopted pedagogical approach from Grammar-Translation, through Audiolingualism to the more recent Communicative Approach.

18.5.1 English Language Teaching Before 1989

In a chronological account of English language teaching (ELT) in Romania before 1989, Presadă and Badea (2010) underline the most important moments of foreign language teaching before the 1990s. Thus, in the early 1920s, the study of foreign languages gained popularity and interest in both private and state schools, when, alongside classical languages, the main European languages began to be studied. This new focus led to the employment of various native language trainers of French,

German, Italian and English, turning Romania into 'a real paradise for teaching and learning foreign languages' (Presadă & Badea, 2010:134).

However, WWII and the following Soviet domination in Romania acted as a deterrent and brought about changes in both the foreign language curriculum and the linguistic hierarchy. Russian became the dominant foreign language at all educational levels, and in 1948, it became compulsory in all schools, an event which triggered a major inflow of specialists in Russian language and a large scale training of Romanian teachers, to prepare them for the newly created demand. In addition to its newly conquered role of fundamental foreign language to be acquired, Russian language teaching also implied numerous changes in the curriculum, the educational system acting as 'the perfect tool for communist propaganda [...]The language textbook dealt with imposed topics and themes. Students had to cope with learning uninteresting, boring materials that emphasised the relationship among three factors, pupil – school – factory, in the light of the supreme ideology, historical materialism, Stalinism [...], while teachers mainly resorted to the old-fashioned Grammar Translation Method. Following the model of the classical languages, teaching was viewed as transmission of a set of grammatical rules, its main criteria being: the medium of instruction was the mother tongue, written exercises helped at practising the target language, translation of written texts into L2 was highly valued, being seen as the supreme aim of foreign language teaching' (Presadă & Badea 2010: 134–135).

Anton Makarenko, who was seen by Aransky and Piskunov (1965:5) as one of the great contributors to pedagogy alongside John Locke, Jean Jacques Rousseau, Johan Pestalozzi, Johann Herbart, Friedrich Deisterweg and K. Ushinsky, based his ideological views on the Marxist teaching, seeing the educational system as embracing all aspects of the student's life and activity. He placed great importance on developing the student's industrial skills (productive labour), the collective ('the school was such a collective, a community of pupils and students headed and directed by the headmaster', with the students being taught to 'coordinate their private interests and aspirations with the interests and aims of the collective') (Aransky & Piskunov, 1965:14,15) and the development of the personality (he promoted the theory and practice of children's self-government, 'which is one of the more effective means of training active and articulate members of society' (1965:17,18). It was Makarenko's belief that the educational system should rely on 'a method – a general, single method, which would, at the same time, give every person a chance to develop his individual traits and preserve his individuality' (1965:19), which naturally translated into a unique method and standardised textbooks. In Romania, these textbooks were in use until well into the 1990s.

Historically, 1965 marks the beginning of a new era, an apparent shift towards a more liberal regime; however, the communist period under N. Ceaușescu was still a totalitarian regime, but one with a newly defined scope, that of building the 'new multilaterally developed society' [my translation]. As Tismăneanu *et al.* claim in the *Report on the Study of the Communist Dictatorship in Romania* [my translation], there was no real change in the Romanian political ideology of the time, no real break from the Stalinist ideology, and according to this report (2006:11). Presadă and Badea support this statement by bringing as arguments the textbooks 'which continued the model of Soviet pedagogy, [with] topics still praising the communist ideology' (2010:137), which were used until alternative textbooks were issued in 1995–1996 for elementary schools and only in 1999 for secondary schools. Describing the new foreign language textbooks of 1970s, Presadă and Badea remark that a change was felt regarding the direction of methodology and pedagogy, and Romanian teachers were exposed to works such as B. F. Skinner's *Scientific Educational Revolution* of 1971, gradually promoting the audiolingual method, when 'lessons were organised so that listening and speaking could be developed by means of a variety of drills and supported by various audio aids. In some cases, the textbooks were even accompanied by vinyl discs to be used by foreign language teachers in the classroom' (Presadă & Badea, 2010:138).

18.5.2 The changes of the 1990s

The 90s brought about significant changes to English language teaching (ELT) in Romania in the context of an expanding EU and the use of English as the new lingua franca; the result was that the 'Ministry of Education increased the number of foreign language classes while lowering the age for starting learning a foreign language to seven years of age, and allotted intensive and bilingual classes in quite a large number of town schools' (Goșa, 2014: 25). Another important factor for the development of ELT was the increased access and exposure to resources in English language teaching methodology and the introduction of alternative textbooks, many of them issued by prestigious publishing houses, such as Cambridge, Oxford or Longman. Most of the new textbooks were written according to the principles of an integrated skills approach and communicative skills were paid more attention, a final rebound after 50 years of standardised textbooks.

As Goșa (2014) confirms, this change was also due to institutions, such as the British Council, the Soros Foundation, USIS, which became more involved with teacher training programmes for the secondary level education, textbook writing and student training. English Language teacher conferences started to be organised and English teachers' associations such RATE (The Romanian Association of Teachers of English) were founded and took an active part in promoting a change towards a Communicative Approach in language teaching; however, it was only in 2000 that local branches such as TETA (Timisoara English Teachers' Association), BETA (Bucharest), CETA (Cluj) and MATE (Moldova) were founded.

As Presadă and Badea assert, an important dimension of the new system was a change of the teacher's mindset, especially with the newly trained professionals.

Admittedly, the old and the new framework co-existed, although there was a certain pressure on the experienced teachers to adjust to the new methods and activities.

'One of the major outcomes was the increased competition among teachers, who could be divided in two categories: traditional teachers, who had to adapt themselves to the new requirements of the curriculum, and the new generation of language trainers, who, formed in the existing context, embraced the communicative dimension of language teaching. This state of affairs resulted from the change of paradigms, which was intrinsic to modern society; language was no longer viewed as a system of rules and structures, but as perpetual communication [...]' (2010:139).

However, the educational system of the mid 90s was still searching for a direction, and although there were numerous attempts to reform it, the results were still short of successful. In a 1998 diagnostic study of the Romanian educational system, Miroiu signals that the reform in education had only affected a small group of 200 teachers from different educational levels, which the author calls '"a closed circle" which was difficult to join' [my translation] (1998:58), this small number illustrating a not very successful reform. Miroiu also remarks on the inculcated mindset that 'the myth of the "gifted teacher" and apostle of the nation still predominates instead of the professional who possesses good working instruments to resort to, regardless of his/her 'natural talent', such as methodologies and training courses' [my translation] (ibid.).

Miroiu's worrying conclusion regarding the philosophy of the Romanian educational system of 1998 was that it 'alienates by its priorities: homework is more important than students, abstract information is more important than the applied knowledge, theoretical-discursive abilities are more important than behavioural competences, information is more important than formation' [my translation] (Miroiu, 1998:65). The examples that support these conclusions include the remarks on the fact that, with a few insignificant exceptions, face-to-face relationships were only built with the teachers; students in secondary and often in tertiary education mostly 'see their colleague's back or profile' for the period of their studies [my translation] (ibid.). Moreover, most communication is directional (teacher–student), of the 'sermon-interrogation' type and communication between students, as well as questions for colleagues are very rare and 'are rather dependent on dissident practices of some nonconformist teachers' [my translation] (ibid.). Miroiu (1998) asserts that a main characteristic of the education system of the time was its tendency towards self-preservation instead of progress and change, its conservatism being rather a state of mind than an ideology, which lay in the hands of those who were afraid of change and one might say that in many respects the same tendency continues to characterise the present day educational system.

18.5.3 ELT in Romania in the 21st century

To the best of my knowledge, this period has not been documented yet and my research reveals that the English language classroom profile is that of mixed practices, which co-occur. However, this blend of approaches, methods and activities is not in the spirit of Kumaravadivelu's (2008) proposed model of the postmethod pedagogy, where the combined methods are suited to tailor the specific needs of the learner, but it is rather the result of the overlying trends and of various inherited practices and mindsets each teacher had been exposed to.

In short, Kumaravadivelu (2008) proposes an entirely new model, which is not based on any particular method or blend, but on three principles and a number of macrostrategies. In fact, he deconstructs the belief that successful language teaching is the result of an exclusive reliance on an approach and argues that the concept of method is surrounded by a number of myths, which need to be acknowledged as such. In his view, at the core of the post-method pedagogical model lie three pedagogic parameters: *particularity* (uniqueness of each language teaching context, built on a critical awareness of local conditions), *practicality* ('involves practicing teachers, either individually or collectively, observing their teaching acts, evaluating their outcomes, identifying problems, finding solutions, and trying them out to see once again what works and what doesn't' (2008:172)), and *possibility* (acknowledges the teacher's and learner's identity and personal ideology as an integral part of the learning process; it implies that the 'experiences participants bring to the pedagogical setting are shaped, not just by what they experience in the classroom, but also by a broader social, economic, and political environment in which they grow up').

However, in the Romanian state-funded schools, the curriculum dictates the content to be taught, but the teaching methods are not imposed on the teacher, so, there is room for a more 'principled approach' (as proposed by post-modern pedagogy). Yet, many of the teachers seem to be stuck in the P-P-P (presentation – practice – production) model of the Audiolingual approach and language classes still focus on building linguistic skills, while the alternative textbooks that are used in schools are aimed at developing language skills in order to communicate effectively in the L2, that is, they deliberately target building a communicative competence.

Although my doctoral research did not focus intentionally on building the profile of the English teacher today, several pieces of information have emerged inadvertently. In an attempt to validate the English language teacher's familiarity with the activities that are aimed at developing speaking skills, I used Finochiarro and Brumfit's index of distinctive characteristics of the Communicative Approach and Audiolingualism (Finocchiaro & Brumfit in Richards and Rodgers, 1986: 66, 67), and I invited two teachers to select the features which best describe what happens/ happened in their classrooms, without labelling the two approaches – teacher C. a retired teacher who was active and very successful throughout the 1990s and teacher R., a young professional, with a 15-year experience in the field, who takes an interest in personal development. Surprisingly, in response to the question 'which of the following statements best describe your teaching', teacher C. favours the principles that are mostly shaped by the Audiolingual Approach, while teacher R. has selected mostly those characteristic of Communicative Language Teaching. Although my choice of the participant teachers rests on convenience, I have deliberately chosen Romanian teachers of a different age, who are genuinely interested in the teaching of English as a representative of a generation, in the belief that the older the teacher, the older the approach to which s/he was exposed; their answers have proven true my assumption.

Moreover, teacher R. was one of the participants in a discussion group on the topic of assessment criteria for the student's performance in an informal debate contest, where most teachers still manifested a strong tendency to assess linguistic performance, rather than communicative Skills, where R. admitted to having difficulty in looking beyond the accuracy of the oral discourse and also confessed to not developing her students' oral skills in the same way as their linguistic skills.

My assertion that underlying mindsets co-occur is best illustrated by the surprising aspects revealed in the discussion group with the 15 secondary school English language teachers I have mentioned above. This group of teachers was preparing an assessment criteria grid for a particular genre of an open debate contest between candidates from different secondary schools in town and since the means of communication during the contest was going to be English, the selection of the participants also included a language test, or proof of at least a B2 level of English proficiency certificate (Cambridge FCE or CAE being the most popular). Yet, during the discussions regarding performance assessment, topics such as linguistic and lexical accuracy and appropriacy came up and teacher B. (an experienced, open minded teacher) admitted to having considered introducing language assessment criteria in the grid, in spite of the preliminary language test for the participants and the communicative nature of the contest. When I asked for an explanation, teacher B. admitted that the grammatical and lexical accuracy are criteria deeply rooted in the mind of language teachers and that they are not trained to positively look at communicative competence in terms of flexibility, naturalness and precision, criteria which according to the Cambridge examinations should characterise the performance of higher level candidates (C1 or C2 according to CEFR). This is not an isolated opinion, it resonates with numerous teachers, who have confessed to finding it difficult to assess speaking skills, as they mentioned in personal talks during teacher training programmes I have been part of.

Additionally, from personal talks with colleagues in the English Language department at the university, who have monitored lessons taught by secondary school teachers, either as part of the student teachers' practicum programme or as members of *Gradul I* examination boards (the highest of the three teacher degrees in Romania), there is a confirmation that there are teachers in the secondary education, who only claim to be using a Communicative Approach in language teaching, but in reality they resort to a blend of activities which belong to the Grammar Translation Approach or to Audiolingualism, more often than not lessons, including translation exercises. A colleague, C., mentioned the example of a secondary school teacher, who had described his approach as communicative in the submitted lesson plan, but the activities consisted of read aloud fragments followed by their translation. Another colleague, M., mentions an instance when the teacher who was demonstrating a communicative approach gave the students a reading task and then added 'but don't worry you'll be asked to read it aloud in a minute'. M. also adds that teachers most often do not make use of the teacher's book and this is also why a textbook, which is built to develop a communicative competence, does not, in fact, achieve this goal.

The profile of the 21st century English teacher in Romania is very different from one case to another and it is not so much due to the teacher's experience; however, the teacher who is successfully embracing a communicative model and assists the learners in developing their language skills so as to be functional in a real-life context, is an isolated case; the majority of teachers follow a combined approach of Grammar Translation and Audiolingualism, although most of the textbooks are written with a focus on language use, rather than language knowledge. What is more, many of these teachers who resist the change of paradigm, knowingly or not, have key roles in the language teacher's community, being trained mentors for future generations of teachers – sometimes, the opposition towards the approach being not a construct of beliefs, but the result of the teacher's exposure to a model of teaching that is viewed as successful in terms of learning experience and which they later reproduce. According to fellow Cambridge examiners, there is only minimal development of speaking skills in class and that is why Romanian candidates often do not achieve the highest results in the speaking component of the main suite Cambridge exams; their problem areas being lower active listening skills, difficulty at interacting naturally with their partner and the underdevelopment of arguments and, sometimes, illogicality.

18.6 Conclusions

From the point of view of the current approaches in Romania, it is clear that while linguistic competence still plays a central role, at least in theory, teachers adopt some type of communicative language teaching, although it is arguable whether the approach is correctly understood and applied. The requirements for the *Definitivat* and *Gradul II* 2013 state exams (the lower of the three teacher degrees) have only recently (MEN, 2008) changed, while up to the year 2000, the English language teaching methodology topics were rather general – 'Theories of foreign language learning' – with often ambiguous phrasing – 'modern methods, procedures and techniques of foreign language learning', – communicative teaching – (my translation – MEN, 2000:5). Firstly, the word 'modern' in Romanian is polysemantic – it may refer to new/recent/updated or belonging to the period referred to as Modernism; with reference to language, a modern language is an actively spoken language, and with reference to an educational system, the semantic overtone of modern is ‘which focuses on disciplines in the field of humanities’ [my translation] (Coteanu, 1998). Secondly, with so many interpretations of the concept of communicative teaching and no specific framing, understanding is left to the teacher, and therefore, it may vary from the traditional communicative language teaching to the postmodern understanding of the communicative intent. The 2008 changes added a certain post-modern component, although, at times, the requirements contain topics that reflect nuances belonging to an older paradigm: ‘Theories of language learning and acquisition. A critical approach’, ‘Critical evaluation of different modern methods, procedures and techniques in teaching a foreign language’, ‘Communicative teaching: principles and types of activities’, ‘Integrating *the linguistic competence(s)* in the English language class’ [my translation, my emphasis] (MEN, 2008:2). The compulsory bibliography has also been slightly revised, but not completely updated (the proposed editions of 1994, 1993, 1989 do not reflect the assertions and the focus of the postmethod pedagogy).

Thus, it is clear that, at present, there is evidence of a certain type of eclecticism in the Romanian ELT pedagogy, which is in keeping with Larsen Freeman’s (2000:177) observation regarding the co-existence of methods in use today, although she admits that ‘they are not equally distributed in classrooms around the world’, but admits that even older approaches, such as the Grammar Translation, have survived for many years and may still be used in schools. However, a line of future investigation is to verify my assertion that this form of eclecticism in Romania is most often not a principled approach, but it rests on the teacher’s personal preference and it is constructed through the accumulation of various individual experiences and information.

Here, we have come full circle to the root of the issue of the discouraged learner, in order to stress again the need for teachers to embrace and adapt to the requirements of a postmodern pedagogy, to adjust to the digital learner’s individual needs, to make use of elements corresponding to various approaches so as to design relevant, engaging learning lessons, where collaborative learning is encouraged, where technology is truly part of everyday learning and teaching, and where learning is continued outside the classroom. Tapscott (2008) admits that old paradigms, embedded in everyday practice in schools, are hard to change, yet he brings numerous examples of schools or programmes, which produced tangible results because of the newly implemented pedagogical practices.

References

Aransky, V., and Piskunov, V. (1965). Preface to A. Makarenko. Problems of Soviet School Education. Translated by O. Shartse. Moscow: Progress Publishers.

Bachman, L., F. (1990). Fundamental Considerations in Language Testing. Oxford: Oxford University Press.

Bachman, L., and Palmer. A. (1996). Language Testing in Practice. Oxford: Oxford University Press.

Canale, M. (1983). From communicative competence to communicative language pedagogy. In J. Richards and R. Schmidt (Eds.), Language Communication (pp. 2-27). London: Longman.

Canale, M., and Swain, M. (1980). Theoretical bases of communicative approaches to second language teaching and testing. Applied Linguistics, 1(1),1-47.

Celce–Murcia, M. (2007). Rethinking the role of Communicative Competence. In E. Alcon Soler, et al. (Eds.), Intercultural Language Use and Language Learning (pp. 41-57). Dordrecht, The Netherlands: Springer.

Celce–Murcia, M., Dörnyei, Z., and Thurrell. S. (1995). A pedagogical framework for communicative competence: A Pedagogically motivated model with content specifications. Issues in Applied Linguistics, 6(2),5-35.

Chomsky, N. (1965). Aspects of the Theory of Syntax. Cambridge: MIT Press.

Comisia Europeană/EACEA/Eurydice. (2012). Dezvoltarea competenţelor cheie în şcolile din Europa: Provocări şi Oportunităţi pentru Politică. Raport Eurydice. Luxemburg: Oficiul pentru Publicaţii al Uniunii Europene. Web, accessed November 2013. <http://eacea.ec.europa.eu/education/eurydice/documents/thematic_reports/145RO.pdf>

Coteanu, I. (coordinator). 1998 (1984). Dicționarul Explicativ al Limbii Române. (2nd ed.). Academia Română, Institutul de Lingvistică Iorgu Iordan. București: Editura Univers Enciclopedic.

Dogioiu, I., E. (2008). Lipsa de comunicare, cea mai gravă boală a societății româneşti? Ziare.com. Web, accessed May 2013. <http://www.ziare.com/politica/stiri-politice/lipsa-de-comunicare-cea-mai-grava-boala-a-societatii-romanesti-362519>

Dumbrăveanu, L. (2006). Comunicarea o problemă a societății românești cu rădăcini în școală? 9AM News. Web, accessed May 2013. < http://www.9am.ro/stiri-revista-presei/Social/29203/Comunicarea-o-problema-a-societatii-romanesti-cu-radacini-in-scoala.html>

Gavrilă, C. (2013). Diplomele IELTS și TOEFL preferate în defavoarea examenelor Cambridge. Portal învățământ. Web, accessed January 2014. <http://www.portalinvatamant.ro/articole/limbi-straine-69/diplomele-ielts-si-toefl-preferate-in-defavoarea-examenului-cambridge-1321.html>.

Goșa, C., M., C. (2014). Diary studies in qualitative research: a tertiary education case study. Szeged:Jate Press.

Hymes, D. (1971). Competence and Performance in linguistic theory. In R. Huxley and E. Ingram. (Eds.), Language Acquisition: Models and Methods (pp3-28). London: Academic Press.

Kumaravadivelu, B. (2008). Understanding Language Teaching. From Method to Postmethod. Taylor &Francis e-library. E-book. Web, accessed May 2014, <https://livelongday.files.wordpress.com/2011/08/kumaraposmethod.pdf.>

Larsen-Freeman, D. 2000 (1986) Techniques and Principles in Language Teaching. (2nd ed.). New York: Oxford University Press.

Ministerul Educației Naționale (MEN) (2000). Limba și literature engleză. Programa pentru examenul de definitivare în învățământ, obţinerea gradului didactic II. Web, accessed May 2013. <file:///C:/Users/valy/Desktop/P_ENGLEZA_def_grII_grI.pdf>

Ministerul Educației Naționale (MEN). (2008). Limba și literatura engleză. Programa pentru obţinerea gradului didactic II. Web, accessed May 2013. <http://www.edu.ro/index.php/programe_perfectiona/c187/>

Miroiu, A. (Ed.), Pasti, V., Ivan, G., Miroiu, M., Marga, A. (1998). Învățământul românesc azi. Studiu de diagnoză. Iași: Polirom.

Miroiu, M. (1998). Filosofia învățământului actual. In A. Miroiu (Ed.), Învățământul românesc azi. Studiu de diagnoză (pp. 51-69). Iași: Polirom.

Mitra, S. (2013). Build a School in the Cloud [Video file]. TED2013. Web, accessed August 2015. <https://www.ted.com/talks/sugata_mitra_build_a_school_in_the_cloud?language=en>

Presadă, D., and Badea, M. (2010). A Brief History of Language Teaching within the Educational System in Romania. Buletinul Universității Petrol - Gaze din Ploiești, Seria Filologie, LXII(4), 133-140.

Richards, J., C., and Rodgers, T., S. (1986). Approaches and Methods in Language Teaching. New York: Cambridge University Press.

Robinson, K. (2008). Changing Paradigms [Video file]. RSA Events, Web, accessed June 2012. <http://www.thersa.org/events/video/archive/sir-ken-robinson>

Solomon, S. (2012). The use of foreign languages by Romanians, under 'EU average'. Radio România Actualităţi, Web, accessed January 2014. <http://www.romania-actualitati.ro/the_use_of_foreign_languages_by_romanians_under_eu_average-42176>

Tapscott, D. (2008). Grown Up Digital. How the Net Generation is Changing Your World. New York: The McGraw-Hill Companies.

Thornbury, S., and Slade, D. (2006). Conversation: From Description to Pedagogy. Cambridge: Cambridge University Press.

Tismăneanu, V. (coordinator). (2006). Raportul Comisiei Prezidenţiale pentru analiza dictaturii comuniste în România. n.p. Web. accessed November 2013. <http://www.presidency.ro/static/ordine/RAPORT_FINAL_CPADCR.pdf>

List of Figures

List of Tables